A Particular Solutions Formula
For Inhomogeneous
Arbitrary Order
Linear Ordinary
Differential Equations

© 2012 Claude Michael Cassano

As I have shown earlier in [6] , [7] , and [8], a particular solution for inhomogeneous linear second, third, and fourth order ordinary differential equations may generally be determined. Applying what was determined from [6] , [7], and [8], and following by example a particular solution formula for arbitrary order is obtained.

The second order particular solution for inhomogeneous linear second order ordinary differential equations may be determined as follows.

Let y_p be a solution to the inhomogeneous ODE:
$$y_p'' + P_0 y_p' + P_1 y_p = W.$$
Then consider:
$$y_p = v y_{h_1} .$$
Differentiating:
$$y_p' = \left[v' + v\left(\frac{y_{h_1}'}{y_{h_1}}\right) \right] y_{h_1} .$$

$$y_p'' = \left[v'' + v'\left(\frac{y_{h_1}'}{y_{h_1}}\right) + v\left(\frac{y_{h_1}'}{y_{h_1}}\right)' \right] y_{h_1} + \left[v' + v\left(\frac{y_{h_1}'}{y_{h_1}}\right) \right] y_{h_1}' .$$

$$= \left[v'' + v'\left(\frac{y_{h_1}'}{y_{h_1}}\right) + v\left(\frac{y_{h_1}'}{y_{h_1}}\right)' \right] y_{h_1} +$$
$$+ \left[v' + v\left(\frac{y_{h_1}'}{y_{h_1}}\right) \right]\left(\frac{y_{h_1}'}{y_{h_1}}\right) y_{h_1}$$

So,
$$W = y_p'' + P_0 y_p' + P_1 y_p$$
$$= \left[v'' + v'\left(\frac{y_{h_1}'}{y_{h_1}}\right) + v\left(\frac{y_{h_1}'}{y_{h_1}}\right)' \right] y_{h_1} +$$
$$+ \left[v' + v\left(\frac{y_{h_1}'}{y_{h_1}}\right) \right]\left(\frac{y_{h_1}'}{y_{h_1}}\right) y_{h_1} + P_0 \left[v' + v\left(\frac{y_{h_1}'}{y_{h_1}}\right) \right] y_{h_1} + P_1 v y_{h_1}$$
$$\Rightarrow W\frac{1}{y_{h_1}} = v'' + \left[2\left(\frac{y_{h_1}'}{y_{h_1}}\right) + P_0 \right] v' + \left[\left(\frac{y_{h_1}'}{y_{h_1}}\right)' + \left(\frac{y_{h_1}'}{y_{h_1}}\right)^2 + P_0\left(\frac{y_{h_1}'}{y_{h_1}}\right) + P_1 \right] v$$

1

$$= v'' + \left[2\left(\frac{y'_{h_1}}{y_{h_1}}\right) + P_0 \right]v' + \left[\frac{y''_{h_1}}{y_{h_1}} + P_0\left(\frac{y'_{h_1}}{y_{h_1}}\right) + P_1 \right]v$$

$$= v'' + \left[2\left(\frac{y'_{h_1}}{y_{h_1}}\right) + P_0 \right]v' + \frac{1}{y_{h_1}}[y''_{h_1} + P_0 y'_{h_1} + P_1 y_{h_1}]v$$

So, if y_{h_1} is a homogeneous solution of the above ODE, then:

$$W\frac{1}{y_{h_1}} = e^{-\int\left[2\left(\frac{y'_{h_1}}{y_{h_1}}\right)+P_0\right]dx}\left[v'e^{\int\left[2\left(\frac{y'_{h_1}}{y_{h_1}}\right)+P_0\right]dx}\right]'$$

$$e^{-\int\left[2\left(\frac{y'_{h_1}}{y_{h_1}}\right)+P_0\right]dx}\int\left(W\frac{1}{y_{h_1}}\right)e^{\int\left[2\left(\frac{y'_{h_1}}{y_{h_1}}\right)+P_0\right]dx}dx = v'$$

$$\Rightarrow y_p = y_{h_1}\left(\int e^{-\int\left[2\left(\frac{y'_{h_1}}{y_{h_1}}\right)+P_0\right]dx}\int\left(W\frac{1}{y_{h_1}}\right)e^{\int\left[2\left(\frac{y'_{h_1}}{y_{h_1}}\right)+P_0\right]dx}dx\right)dx$$

$$= y_{h_1}\left(\int\frac{1}{y_{h_1}^2}e^{-\int P_0 dx}\int\left[\left(W\frac{1}{y_{h_1}}\right)y_{h_1}^2 e^{\int P_0 dx}\right]dx\right)dx$$

$$y_p = y_{h_1}\left(\int\frac{1}{y_{h_1}^2}e^{-\int P_0 dx}\int\left[\left(W\frac{1}{y_{h_1}}\right)y_{h_1}^2 e^{\int P_0 dx}\right]dx\right)dx.,$$

is a particular solution for any inhomogeneous linear second order ordinary differential equation.

As an example, taken from Zill's text 2nd ed.:

$y'' - 3y' = 8e3x + 4\sin x$

$\Rightarrow y_p = (8/3)e^{3x} + (6/5)\cos x - (2/5)\sin x$

The above formula gives:

choosing: $y_h = 1$

$\Rightarrow y_p = 1\int[(1/1^2)\int(8e^{3x} + 4\sin x)1e^{\int(-3)dx}dx]e^{-\int(-3)dx}dx$

Integrating by parts, or going to online integral tables, like:

http://www.sosmath.com/tables/integral/integ27/integ27.html

We get:

$yp = (8/3)xe^{3x} - (8/9)e^{3x} + k_1 e^{3x} + k_2 +$
$\quad + 4\int\{e^{-3x}(-3\sin x - \cos x)/(9+1) + k_3\}e^{3x}dx$

which is clearly the same, except for the homogeneous solution parts (which can be absorbed into the homogeneous solution).

Another example, again taken from Zill's text 2nd ed.:

$$y'' - y = 1/x$$
$$\Rightarrow y_p = (1/2)e^x \int^x t^{-1}e^{-t}dt - (1/2)e^{-x}\int^x t^{-1}e^t dt$$

The above formula gives:

choosing: $y_h = e^x$

$$\Rightarrow y_p = e^x \int^x [(1/e^{2t})\int^t (1/s)e^s e^{\int(0)ds}ds]e^{-\int(0)dt}dt$$
$$= e^x \int^x [e^{-2t}\int^t s^{-1}e^s ds]dt$$

That the difference between these two is a constant times e^x is easily seen by dividing each by e^x and differentiating each. Since that is a homogeneous solution, and so is the first; then we again see corroboration.

As a more general example, consider the linear inhomogeneous ODEs with constant coefficients:

$$y'' + Ay' + By = Ce^{ax} + De^{bx}.$$
$$y_h = e^{mx} \text{ , where } m \text{ is a root of the auxiliary equation.}$$

So,

$$y_p = e^{mx}\int e^{-2mx}(\int (Ce^{ax} + De^{bx})e^{(m+A)x}dx)e^{-Ax}dx.$$
$$\Rightarrow y_p = e^{mx}\int (\int Ce^{(a+m+A)x} + De^{(b+m+A)x}dx)e^{-(2m+A)x}dx.$$
$$\Rightarrow y_p = e^{mx}\int \left(\frac{C}{a+m+A}e^{(a+m+A)x} + \frac{D}{b+m+A}e^{(b+m+A)x}\right)e^{-(2m+A)x}dx, (a+m+A \neq 0)$$
$$\Rightarrow y_p = e^{mx}\int \left(\frac{C}{a+m+A}e^{(a+m+A-2m-A)x} + \frac{D}{b+m+A}e^{(b+m+A-2m-A)x}\right)dx.$$
$$\Rightarrow y_p = e^{mx}\int \left(\frac{C}{a+m+A}e^{(a-m)x} + \frac{D}{b+m+A}e^{(b-m)x}\right)dx.$$
$$\Rightarrow y_p = e^{mx}\left(\frac{C}{(a+m+A)(a-m)}e^{(a-m)x} + \frac{D}{(b+m+A)(b-m)}e^{(b-m)x}\right),$$
$$(a+m+A \neq 0, a-m \neq 0)$$

$$\Rightarrow y_p = Re^{ax} + Se^{bx},$$

where :

$$R = \begin{cases} \dfrac{C}{(a+m+A)(a-m)}, & (a+m+A \neq 0, a-m \neq 0) \\[2ex] \dfrac{C}{(a+m+A)}\cdot x, & (a+m+A \neq 0, a-m = 0) \\[2ex] \dfrac{C}{(a-m)}(x+1), & (a+m+A = 0, a-m \neq 0) \\[2ex] \dfrac{C}{2}x^2, & (a+m+A = 0, a-m = 0) \end{cases}$$

$$S = \begin{cases} \dfrac{D}{(b+m+A)(a-m)}, & (b+m+A \neq 0, b-m \neq 0) \\[2ex] \dfrac{D}{(b+m+A)}\cdot x, & (b+m+A \neq 0, b-m = 0) \\[2ex] \dfrac{D}{(b-m)}(x+1), & (b+m+A = 0, b-m \neq 0) \\[2ex] \dfrac{D}{2}x^2, & (b+m+A = 0, b-m = 0) \end{cases}$$

(*note* : $a + m + A = 0 \Rightarrow m - a = -A - a - a = -A - 2a = -A + 2m + 2A = 2m + A$)

I believe it is clear, without going into detail, that:

$$y'' + Ay' + By = \sum_{j=1}^{N} C_j e^{a_j x} \Rightarrow y_p = \sum_{j=1}^{N} S_j e^{a_j x},$$

$$\text{where} : S_j = \begin{cases} \dfrac{C_j}{(a_j + m + A)(a_j - m)} & ,(a_j + m + A \neq 0, a_j - m \neq 0) \\[3mm] \dfrac{C_j}{(a_j + m + A)} \cdot x & ,(a_j + m + A \neq 0, a_j - m = 0) \\[3mm] \dfrac{C_j}{(a_j - m)}(x+1) & ,(a_j + m + A = 0, a_j - m \neq 0) \\[3mm] \dfrac{C_j}{2}x^2 & ,(a_j + m + A = 0, a_j - m = 0) \end{cases}$$

As noted above, applying the same
method to the inhomogeneous linear
third order ordinary differential equation.

Consider the homogeneous linear third order ordinary differential equation.
As before, consider:
$$y = v y_{h_1} \ .$$
Differentiating:

$$y' = \left[v' + v\left(\frac{y'_{h_1}}{y_{h_1}} \right) \right] y_{h_1} \ .$$

$$y'' = \left[v'' + v'\left(\frac{y'_{h_1}}{y_{h_1}} \right) + v\left(\frac{y'_{h_1}}{y_{h_1}} \right)' \right] y_{h_1} + \left[v' + v\left(\frac{y'_{h_1}}{y_{h_1}} \right) \right] y'_{h_1} \ .$$

$$= \left[v'' + v'\left(\frac{y'_{h_1}}{y_{h_1}} \right) + v\left(\frac{y'_{h_1}}{y_{h_1}} \right)' \right] y_{h_1} +$$
$$+ \left[v' + v\left(\frac{y'_{h_1}}{y_{h_1}} \right) \right]\left(\frac{y'_{h_1}}{y_{h_1}} \right) y_{h_1}$$

$$y''' = \left[v'' + v'\left(\frac{y'_{h_1}}{y_{h_1}} \right) + v\left(\frac{y'_{h_1}}{y_{h_1}} \right)' \right]' y_{h_1} +$$
$$+ \left[v'' + v'\left(\frac{y'_{h_1}}{y_{h_1}} \right) + v\left(\frac{y'_{h_1}}{y_{h_1}} \right)' \right] y'_{h_1} +$$
$$+ \left(\left[v' + v\left(\frac{y'_{h_1}}{y_{h_1}} \right) \right]\left(\frac{y'_{h_1}}{y_{h_1}} \right) \right)' y_{h_1} +$$
$$+ \left[v' + v\left(\frac{y'_{h_1}}{y_{h_1}} \right) \right]\left(\frac{y'_{h_1}}{y_{h_1}} \right) y'_{h_1}$$

For brevity, let: $s = \dfrac{y'_{h_1}}{y_{h_1}}$

then:

$$y' = (v' + vs)y_{h_1} \ .$$
$$y'' = [v'' + v's + vs']y_{h_1} + [v' + vs]y'_{h_1} \ .$$
$$= (v'' + 2v's + vs' + vs^2)y_{h_1}$$
$$y''' = [v'' + v's + vs']'y_{h_1} + [v'' + v's + vs']y'_{h_1} +$$
$$+ ([v' + vs]s)'y_{h_1} + [v' + vs]sy'_{h_1}$$
$$= (v''' + 3v''s + 3v's' + vs'' + 3v's^2 + 3vss' + vs^3)y_{h_1}$$

Note: if $v = 1$:

$$y = y_{h_1} \ .$$
$$y' = sy_{h_1} \ .$$
$$y'' = (s' + s^2)y_{h_1}$$
$$y''' = (s'' + 3s's + s^3)y_{h_1}.$$

So:

$$y''' + P_0 y'' + P_1 y' =$$
$$= y'''_{h_1} + P_0 y''_{h_1} + P_1 y'_{h_1}$$
$$= y_{h_1}[s'' + 3s's + s^3 + P_0(s' + s^2) + P_1 s]$$

So, if:

$$y'''_{h_1} + P_0 y''_{h_1} + P_1 y'_{h_1} + P_2 y_{h_1} = 0$$
then there exists a P_2 such that:
$$P_2 = -s'' - 3s's - s^3 - P_0(s' + s^2) - P_1 s$$

Returning to general v :

$$y''' + P_0 y'' + P_1 y' + P_2 y =$$
$$= y_{h_1}[v''' + 3v''s + 3v's' + vs'' + 3v's^2 + 3vss' + vs^3 +$$
$$+ P_0(v'' + 2v's + vs' + vs^2) + P_1(v' + vs) + P_2 v]$$
$$= y_{h_1}[v''' + 3v''s + 3v's' + 3v's^2 +$$
$$+ P_0(v'' + 2v's) + P_1 v' + P_0 v +$$
$$+ \{s'' + 3s's + s^3 + P_0(s' + s^2) + P_1 s + P_2\}v]$$
$$= y_{h_1}[v''' + 3v''s + 3v's' + 3v's^2 + P_0(v'' + 2v's) + P_1 v']$$
$$= y_{h_1}[(v')'' + (v')'(3s + P_0) + (v')(3s' + 3s^2 + 2P_0 s + P_1)]$$

So, if:

$$y''' + P_0 y'' + P_1 y' + P_2 y = W$$

then:

$$W\frac{1}{y_{h_1}} = (v')'' + (v')'(3s + P_0) + (v')(3s' + 3s^2 + 2P_0 s + P_1)$$

is an inhomogeneous linear second order
ordinary differential equation (in v').

And, $e^{\int sdx} = y_{h_1}$ is a homogeneous solution to
$$y''' + P_0 y'' + P_1 y' + P_0 y = 0$$
as defined.

So, applying the above formula for the
particular solution of this inhomogeneous

linear second order ordinary differential equation.

$$v' = u_p = u_h \int \frac{1}{u_h^2}\left[\int\left(We^{-\int sdx}\frac{1}{u_h}\right)u_h^2 e^{\int(3s+P_0)dx}dx\right]e^{-\int(3s+P_0)dx}dx.$$

where u_h is a homogeneous solution of:
$$u_h'' + (3s + P_0)u_h' + (3s' + 3s^2 + 2P_0 s + P_1)u_h = 0.$$

Now, let a particular solution be denoted:
$$y_p = ve^{\int sdx} = vy_{h_1} \Rightarrow u_p = v' = \left(\frac{y_p}{y_{h_1}}\right)'$$

And, recalling the process of reduction of order.

For
$$y_{h_2} = wy_{h_1} \Rightarrow w = \frac{y_{h_2}}{y_{h_1}}$$

then:
$$y_{h_2}' = wy_{h_1}' + w'y_{h_1}$$
$$y_{h_2}'' = wy_{h_1}'' + 2w'y_{h_1}' + w''y_{h_1}$$
$$y_{h_2}''' = wy_{h_1}''' + 3w'y_{h_1}'' + 3w''y_{h_1}' + w'''y_{h_1}$$

So:
$$0 = y_{h_2}''' + P_0 y_{h_2}'' + P_1 y_{h_2}' + P_0 y_{h_2} =$$
$$= wy_{h_1}''' + 3w'y_{h_1}'' + 3w''y_{h_1}' + w'''y_{h_1} +$$
$$+ P_0(wy_{h_1}'' + 2w'y_{h_1}' + w''y_{h_1}) +$$
$$+ P_1(wy_{h_1}' + w'y_{h_1}) + P_0 wy_{h_1}$$
$$= w(y_{h_1}''' + P_0 y_{h_1}'' + P_1 y_{h_1}' + P_0 y_{h_1}) +$$
$$+ 3w'y_{h_1}'' + 3w''y_{h_1}' + w'''y_{h_1} +$$
$$+ P_0(2w'y_{h_1}' + w''y_{h_1}) + P_1 w'y_{h_1}$$
$$= w'''y_{h_1} + w''(3y_{h_1}' + P_0 y_{h_1}) +$$
$$+ w'(3y_{h_1}'' + 2P_0 y_{h_1}' + P_1 y_{h_1})$$
$$= y_{h_1}\left[w''' + w''\left(3\frac{y_{h_1}'}{y_{h_1}} + P_0\right) + \right.$$
$$\left. + w'\left(3\frac{y_{h_1}''}{y_{h_1}} + 2P_0\frac{y_{h_1}'}{y_{h_1}} + P_1\right)\right]$$
$$= w''' + w''\left(3\frac{y_{h_1}'}{y_{h_1}} + P_0\right) + w'\left(3\frac{y_{h_1}''}{y_{h_1}} + 2P_0\frac{y_{h_1}'}{y_{h_1}} + P_1\right)$$
$$= (w')'' + (w')'\left(3\frac{y_{h_1}'}{y_{h_1}} + P_0\right) + (w')\left(3\left(\frac{y_{h_1}'}{y_{h_1}}\right)' + 3\left(\frac{y_{h_1}'}{y_{h_1}}\right)^2 + 2P_0\frac{y_{h_1}'}{y_{h_1}} + P\right)$$

But: $y_{h_1} = e^{\int sdx} \Rightarrow \frac{y_{h_1}'}{y_{h_1}} = s$

So, $w' = \left(\frac{y_{h_2}}{y_{h_1}}\right)'$ is a homogeneous solution to:
$$u_h'' + (3s + P_0)u_h' + (3s' + 3s^2 + 2P_0 s + P_1)u_h = 0.$$

And, therefore:
$$\left(\frac{y_p}{y_{h_1}}\right)' = v' = u_p$$

$$= \left(\frac{y_{h_2}}{y_{h_1}}\right)' \int \frac{1}{\left[\left(\frac{y_{h_2}}{y_{h_1}}\right)'\right]^2} \left[\int \left(W\frac{1}{y_{h_1}} \frac{1}{\left(\frac{y_{h_2}}{y_{h_1}}\right)'}\right) \left[\left(\frac{y_{h_2}}{y_{h_1}}\right)'\right]^2 e^{\int \left[3\left(\frac{y'_{h_1}}{y_{h_1}}\right)+P_0\right]dx} dx\right] \times$$

$$\times e^{-\int \left[\left(3\left(\frac{y'_{h_1}}{y_{h_1}}\right)+P_0\right)\right]dx} dx.$$

So, finally.

$$y_p = y_{h_1} \int \left(\frac{y_{h_2}}{y_{h_1}}\right)' \left\{\int \frac{1}{\left[\left(\frac{y_{h_2}}{y_{h_1}}\right)'\right]^2} \left(\int \left(W\frac{1}{y_{h_1}} \frac{1}{\left(\frac{y_{h_2}}{y_{h_1}}\right)'}\right) \left[\left(\frac{y_{h_2}}{y_{h_1}}\right)'\right]^2 y_{h_1}^3 e^{\int P_0 dx} dx\right) \times\right.$$

$$\left. \times y_{h_1}^{-3} e^{-\int P_0 dx} dx\right\} dx$$

$$\boxed{y_p = y_{h_1} \int \left(\frac{y_{h_2}}{y_{h_1}}\right)' \left\{\int \frac{1}{y_{h_1}^3} \frac{1}{\left[\left(\frac{y_{h_2}}{y_{h_1}}\right)'\right]^2} \left(\int \left(W\frac{1}{y_{h_1}} \frac{1}{\left(\frac{y_{h_2}}{y_{h_1}}\right)'}\right) y_{h_1}^3 \left[\left(\frac{y_{h_2}}{y_{h_1}}\right)'\right]^2 e^{\int P_0 dx} dx\right) \times\right.}$$

$$\boxed{\left. \times e^{-\int P_0 dx} dx\right\} dx,}$$

is a particular solution for any inhomogeneous linear third order ordinary differential equation.

Example 1, consider from [2], Exercise 6.5, 3:

$z''' + 3z'' - 4z = e^{2x} \Rightarrow z_p = \frac{1}{16}e^{2x}$.

$m^3 + 3m^2 - 4 = 0$,

By inspection, 1 is a root. So, by long division:

$(m-1)(m^2 + 4m + 4) = (m-1)(m+2)^2 = m^3 + 3m^2 - 4$

So, roots are: 1 and -2

$P_0 = 3$, $W = e^{2x}$

let: $z_{h_1} = e^x$, $z_{h_2} = e^{-2x} \Rightarrow \left(\frac{z_{h_2}}{z_{h_1}}\right)' = -3e^{-3x}$

$\Rightarrow z_p = e^x \int \left\{(-3e^{-3x}) \int (\frac{1}{9}e^{6x}) e^{-3x} (\int e^{2x} e^{2x} (-3e^{-3x}) e^{3x} dx) e^{-3x} dx\right\} dx$

$= e^x \int \{e^{-3x} \int e^{3x} (\int e^{4x} dx) e^{-3x} dx\} dx = e^x \int \left\{e^{-3x} \int \left(\frac{e^{4x}}{4}\right) dx\right\} dx$

$= \frac{1}{4} e^x \int \left\{e^{-3x} \left(\frac{e^{4x}}{4}\right)\right\} dx = \frac{1}{16} e^x \int \{e^x dx\} = \frac{1}{16} e^x e^x = \frac{1}{16} e^{2x}$

Example 2, consider from [5], Exercise 4.5, 31:

$y''' - 2y'' - y' + 2y = e^{3x} \Rightarrow y_p = \frac{1}{8}e^{3x}$.

$m \in \{-1, 1, 2\}$, $P_0 = -2$, $W = e^{3x}$

for simplicity, let: $y_{h_1} = e^x$, $y_{h_2} = e^{2x} \Rightarrow \left(\frac{y_{h_2}}{y_{h_1}}\right)' = e^x$

$$\Rightarrow y_p = e^x \int e^x \{\int e^{-2x} e^{-3x} (\int e^{3x} e^{2x} e^x e^{-2x} dx) e^{2x} dx\} dx$$
$$= e^x \int e^x \{\int e^{-3x} (\int e^{4x} dx) dx\} dx = e^x \int e^x \{\int e^{-3x} \left(\frac{e^{4x}}{4}\right) dx\} dx$$
$$= \tfrac{1}{4} e^x \int e^x \{\int e^x dx\} dx = \tfrac{1}{4} e^x \int e^x e^x dx = \tfrac{1}{4} e^x \int e^{2x} dx$$
$$= \tfrac{1}{4} e^x \left(\frac{e^{2x}}{2}\right) = \tfrac{1}{8} e^{3x}$$

Example 3, consider from [2], Exercise 6.4, 7:

$y''' + 3y'' - 4y = e^{-2x} \Rightarrow y_p = -\tfrac{1}{6} x^2 e^{-2x}$.

By inspection: 1 and -2 are roots of the auxiliary equation.

$P_0 = 3$, $W = e^{-2x}$

let: $y_{h_1} = e^x$, $y_{h_2} = e^{-2x} \Rightarrow \left(\frac{y_{h_2}}{y_{h_1}}\right)' = -3e^{-3x}$

$$\Rightarrow y_p = e^x \int \{(-3e^{-3x}) \int (\tfrac{1}{9} e^{6x}) e^{-3x} (\int e^{-2x} e^{2x} (-3e^{-3x}) e^{3x} dx) e^{-3x} dx\} dx$$
$$= e^x \int \{e^{-3x} \int e^{3x} (\int dx) e^{-3x} dx\} dx = e^x \int \{e^{-3x} \int x dx\} dx$$
$$= e^x \int \{e^{-3x} \frac{x^2}{2}\} dx = \tfrac{1}{2} e^x \int x^2 e^{-3x} dx = \tfrac{1}{2} e^x e^{-3x} \left(\frac{x^2}{-3} - \frac{2x}{9} + \frac{2}{-27}\right)$$
$$= -\tfrac{1}{6} x^2 e^{-2x} - \tfrac{2}{9} x e^{-2x} - \tfrac{2}{27} e^{-2x}$$

The 1st term corresponds and I already double checked it as a particular solution. The last term is a homogeneous solution, so only the middle term needs to be checked, and it, too, is a homogeneous solution(because -2 is a double root).

Example 4, consider from [2], Exercise 6.5, 1:

$y''' - 3y'' + 4y = e^{2x}$.

for verification.

$A = -3$, $D = 1$, $a = 2$

$m^3 - 3m^2 + 4 = 0$,

By inspection, -1 is a root. So, by long division:

$(m+1)(m^2 - 4m + 4) = (m+1)(m-2)^2 = m^3 - 3m^2 + 4$

So, roots are: -1 and $+2$

And, so, two distinct homogeneous solutions are: $y_{h_1} = e^x, y_{h_2} = e^{-2x}$

$(a + m_1 + m_2 + A = 2 - 1 + 2 - 3 = 0$,

$a - m_1 = 2 + 1 \neq 0, a - m_2 = 2 - 2 = 0)$

$$\Rightarrow y_p = \tfrac{1}{2} e^{2x} \left[\frac{x^2}{2 - (-1)} - \frac{2x}{(3)^2} + \frac{2}{(3)^3}\right] = \tfrac{1}{2} e^{2x} \left(\tfrac{1}{3} x^2 - \tfrac{2}{9} x + \tfrac{2}{81}\right)$$

Reference [2] gives $\tfrac{1}{6} x^2 e^{2x}$ as particular solution, and it is, but so is the above, because 2 is a double root of the auxiliary equation.

As a more general example, consider the linear inhomogeneous ODEs with constant coefficients:

$$y''' + Ay'' + By' + Cy = De^{ax}$$
$$y_{h_1} = e^{m_1 x}, y_{h_2} = e^{m_2 x} \text{ , where } m_1 \text{ , } m_2 \text{ are}$$

distinct roots of the auxiliary equation.

So,

$$\left(\frac{y_{h_2}}{y_{h_1}}\right)' = (m_2 - m_1)e^{(m_2-m_1)x} \text{ , } P_2 = A \text{ , } W = De^{ax}$$

$$y_p = e^{m_1 x} \times$$

$$\times \int \left(\frac{y_{h_2}}{y_{h_1}}\right)' \left\{ \int \frac{1}{\left[\left(\frac{y_{h_2}}{y_{h_1}}\right)'\right]^2} \frac{1}{(e^{m_1 x})^3} \left(\int De^{ax}(e^{m_1 x})^2 (m_2 - m_1)e^{(m_2-m_1)x}e^{Ax}dx \right)e^{-Ax}dx \right\}$$

$$y_p = e^{m_1 x} \int \left(\frac{y_{h_2}}{y_{h_1}}\right)' \left\{ \int \frac{1}{\left[\left(\frac{y_{h_2}}{y_{h_1}}\right)'\right]^2} \frac{1}{(e^{m_1 x})^3} ((m_2 - m_1)D \int e^{(a+2m_1+m_2-m_1+a)x}dx)e^{-Ax}dx \right\}$$

$$y_p = e^{m_1 x} \int \left(\frac{y_{h_2}}{y_{h_1}}\right)' \left\{ \int \frac{(m_2 - m_1)}{(m_2 - m_1)^2} e^{-2(m_2-m_1)x}e^{-3m_1 x} \left(D\frac{e^{(a+m_1+m_2+A)x}}{a+m_1+m_2+A} \right)e^{-Ax}dx \right\}dx,$$

$$y_p = e^{m_1 x} \int \left(\frac{y_{h_2}}{y_{h_1}}\right)' \left\{ \frac{D}{(m_2 - m_1)(a+m_1+m_2+A)} \int e^{[-2(m_2-m_1)-3m_1+(a+m_1+m_2+A)-A]x}dx \right\}dx$$

$$y_p = e^{m_1 x} \int (m_2 - m_1)e^{(m_2-m_1)x} \left\{ \frac{D}{(m_2 - m_1)(a+m_1+m_2+A)} \int e^{(-m_2+a)x}dx \right\}dx,$$

$$y_p = e^{m_1 x} \int e^{(m_2-m_1)x} \frac{D}{(a+m_1+m_2+A)(-m_2+a)} e^{(-m_2+a)x}dx,$$

$$y_p = \frac{D}{(a+m_1+m_2+A)(-m_2+a)} e^{m_1 x} \int e^{(a-m_1)x}dx,$$

$$y_p = \frac{D}{(a+m_1+m_2+A)(a-m_2)(a-m_1)} e^{ax},$$

$$(a+m_1+m_2+A \neq 0, a-m_1 \neq 0, a-m_2 \neq 0).$$

$(a+m_1+m_2+A = 0):$ $(\Rightarrow -A = a+m_1+m_2)$

$$y_p = e^{m_1 x} \int \left(\frac{y_{h_2}}{y_{h_1}}\right)' \left\{ \int \frac{(m_2 - m_1)}{(m_2 - m_1)^2} e^{-2(m_2-m_1)x} \frac{1}{(e^{m_1 x})^3} Dxe^{-Ax}dx \right\}dx,$$

$$y_p = e^{m_1 x} \int (m_2 - m_1)e^{(m_2+m_1)x} \left\{ \int \frac{(m_2 - m_1)}{(m_2 - m_1)^2} e^{-2(m_2-m_1)x} \frac{1}{(e^{m_1 x})^3} Dxe^{-Ax}dx \right\}dx,$$

$$y_p = e^{m_1 x}D \int e^{(m_2-m_1)x} \{ \int e^{(-2m_2+2m_1-3m_1-A)x}xdx \}dx,$$

$$y_p = e^{m_1 x}D \int e^{(m_2-m_1)x} \{ \int e^{(-2m_2+2m_1-3m_1+a+m_1+m_2)x}xdx \}dx,$$

$$y_p = e^{m_1 x}D \int e^{(m_2-m_1)x} \{ \int e^{(-m_2+a)x}xdx \}dx,$$

$(a-m_2 = 0):$

$$y_p = e^{m_1 x}D \int e^{(m_2-m_1)x} \frac{1}{2}x^2dx,$$

$$y_p = e^{m_1 x}\frac{1}{2}D \int e^{(a-m_1)x}x^2dx,$$

$$y_p = \frac{1}{2}De^{ax}\left[\frac{x^2}{a-m_1} - \frac{2x}{(a-m_1)^2} + \frac{2}{(a-m_1)^3}\right],$$

$$(a+m_1+m_2+A=0, a-m_1 \neq 0, a-m_2=0).$$

The above process may be economized by re-using the analysis for 0 & y_{h_2} as was used for W & y_p, but for clarity in this first iteration each was isolated and treated separately.

This process may be applied to the fourth order, then the fifth, and so on, yielding an algorithm for a particular solution for arbitrary order, as follows.

This time, starting with a general n-th order approach along the same lines as the general reduction of order analysis.

Let: $P_{-1} \equiv 1$,
and:
$$\sum_{m=0}^{n} P_{n-1-m}y^{(m)} = W$$

is a nth order ODE.

Let $y = y_0 v$

$$\Rightarrow y^{(m)} = \sum_{k=0}^{m}\binom{m}{k}v^{(m-k)}y_0^{(k)} = \sum_{k=0}^{m}\binom{m}{k}v^{(m)}y_0^{(m-k)}.$$

$\therefore$

$$\sum_{m=0}^{n} P_{n-1-m}y^{(m)} = \sum_{m=0}^{n} P_{n-1-m}\left[\sum_{k=0}^{m}\binom{m}{k}v^{(m-k)}y_0^{(k)}\right]$$

$$= \sum_{m=0}^{n} y_0^{(m)}\left[\sum_{k=m}^{n}\binom{k}{m}P_{m-1-k}v^{(k)}\right]$$

$$= \sum_{m=0}^{n} v^{(m)}\left[\sum_{k=0}^{n-m}\binom{n-k}{n-m-k}P_{k-1}y_0^{(n-m-k)}\right].$$

So:

$$\sum_{m=0}^{n} P_{n-1-m}y^{(m)} = \sum_{k=0}^{n}\binom{n-k}{n-k}P_{k-1}y^{(n-k)}$$

$$= \sum_{m=0}^{n} v^{(m)}\left[\sum_{k=0}^{n-m}\binom{n-k}{n-m-k}P_{k-1}y_0^{(n-m-k)}\right]$$

$$= v^{(0)}\left[\sum_{k=0}^{n-0}\binom{n-k}{n-0-k}P_{k-1}y_0^{(n-0-k)}\right] +$$

$$+ \sum_{m=1}^{n} v^{(m)}\left[\sum_{k=0}^{n-m}\binom{n-k}{n-m-k}P_{k-1}y_0^{(n-m-k)}\right]$$

$$= v^{(0)}\left[\sum_{k=0}^{n}P_{k-1}y_0^{(n-k)}\right] +$$

$$+ \sum_{m=1}^{n} v^{(m)}\left[\sum_{k=0}^{n-m}\binom{n-k}{n-m-k}P_{k-1}y_0^{(n-m-k)}\right]$$

Then, for y_0 such that: $\sum_{k=0}^{n} P_{k-1} y_0^{(n-k)} = 0$:

$$\sum_{m=0}^{n} P_{n-1-m} y^{(m)} = \sum_{m=1}^{n} v^{(m)} \left[\sum_{k=0}^{n-m} \binom{n-k}{n-m-k} P_{k-1} y_0^{(n-m-k)} \right]$$

$v' = w$:

$$\Rightarrow W = \sum_{m=1}^{n} w^{(m-1)} \left[\sum_{k=0}^{n-m} \binom{n-k}{n-m-k} P_{k-1} y_0^{(n-m-k)} \right]$$

As noted earlier, taking the ordered pair $(W, y_0) = \left(0, y_{h_1} = e^{\int s\,dx} \Rightarrow s = \dfrac{y'_{h_1}}{y_{h_1}} \right)$:

$$\Rightarrow 0 = \sum_{m=1}^{n} w_\mu^{(m-1)} \left[\sum_{k=0}^{n-m} \binom{n-k}{n-m-k} P_{k-1} y_{h_1}^{(n-m-k)} \right]$$

$$= \sum_{m=1}^{n} w_\mu^{(m-1)} \left[\sum_{k=0}^{n-m} \binom{n-k}{n-m-k} P_{k-1} \frac{y_{h_1}^{(n-m-k)}}{y_{h_1}} \right]$$

$$= w_\mu^{(n-1)} \left[\sum_{k=0}^{n-n} \binom{n-k}{n-n-k} P_{k-1} \frac{y_{h_1}^{(n-n-k)}}{y_{h_1}} \right] +$$

$$+ w_\mu^{(n-1-1)} \left[\sum_{k=0}^{n-(n-1)} \binom{n-k}{n-(n-1)-k} P_{k-1} \frac{y_{h_1}^{(n-(n-1)-k)}}{y_{h_1}} \right] +$$

$$+ \sum_{m=1}^{n-2} w_\mu^{(m-1)} \left[\sum_{k=0}^{n-m} \binom{n-k}{n-m-k} P_{k-1} \frac{y_{h_1}^{(n-m-k)}}{y_{h_1}} \right]$$

$$= w_\mu^{(n-1)} \left[\binom{0}{0} P_{0-1} \frac{y_{h_1}^{(0)}}{y_{h_1}} \right] +$$

$$+ w_\mu^{(n-1-1)} \left[\sum_{k=0}^{1} \binom{n-k}{1-k} P_{k-1} \frac{y_{h_1}^{(1-k)}}{y_{h_1}} \right] +$$

$$+ \sum_{m=1}^{n-2} w_\mu^{(m-1)} \left[\sum_{k=0}^{n-m} \binom{n-k}{n-m-k} P_{k-1} \frac{y_{h_1}^{(n-m-k)}}{y_{h_1}} \right]$$

$$= w_\mu^{(n-1)} P_{-1} +$$

$$+ w_\mu^{(n-1-1)} \left[\binom{n-0}{1-0} P_{0-1} \frac{y_{h_1}^{(1-0)}}{y_{h_1}} + \binom{n-1}{1-1} P_{1-1} \frac{y_{h_1}^{(1-1)}}{y_{h_1}} \right] +$$

$$+ \sum_{m=1}^{n-2} w_\mu^{(m-1)} \left[\sum_{k=0}^{n-m} \binom{n-k}{n-m-k} P_{k-1} \frac{y_{h_1}^{(n-m-k)}}{y_{h_1}} \right]$$

$$= w_\mu^{(n-1)} P_{-1} +$$

$$+ w_\mu^{(n-2)} \left[n P_{-1} \frac{y_{h_1}^{(1)}}{y_{h_1}} + \binom{n-1}{0} P_0 \frac{y_{h_1}^{(0)}}{y_{h_1}} \right] +$$

$$+ \sum_{m=1}^{n-2} w_\mu^{(m-1)} \left[\sum_{k=0}^{n-m} \binom{n-k}{n-m-k} P_{k-1} \frac{y_{h_1}^{(n-m-k)}}{y_{h_1}} \right]$$

$$= w_\mu^{(n-1)} + w_\mu^{(n-2)} \left[n \frac{y'_{h_1}}{y_{h_1}} + P_0 \right] +$$

$$+ \sum_{m=1}^{n-2} w_\mu^{(m-1)} \left[\sum_{k=0}^{n-m} \binom{n-k}{n-m-k} P_{k-1} \frac{y_{h_1}^{(n-m-k)}}{y_{h_1}} \right]$$

This is a homogeneous $(n-1)-th$ order linear ODE in
$$w_\mu = v'_\mu = \left(\frac{y_{h_\mu}}{y_{h_1}} \right)', (y_{h_\mu} \neq y_{h_1}).$$

And, now, taking the ordered pair $(W, y_0) = (W, y_p)$:

$$\Rightarrow W = \sum_{m=1}^{n} w_p^{(m-1)} \left[\sum_{k=0}^{n-m} \binom{n-k}{n-m-k} P_{k-1} y_{h_1}^{(n-m-k)} \right]$$

$$\Rightarrow W \frac{1}{y_{h_1}} = \sum_{m=1}^{n} w_p^{(m-1)} \left[\sum_{k=0}^{n-m} \binom{n-k}{n-m-k} P_{k-1} \frac{y_{h_1}^{(n-m-k)}}{y_{h_1}} \right]$$

$$= w_p^{(n-1)} \left[\sum_{k=0}^{n-n} \binom{n-k}{n-n-k} P_{k-1} \frac{y_{h_1}^{(n-n-k)}}{y_{h_1}} \right] +$$

$$+ w_p^{(n-1-1)} \left[\sum_{k=0}^{n-(n-1)} \binom{n-k}{n-(n-1)-k} P_{k-1} \frac{y_{h_1}^{(n-(n-1)-k)}}{y_{h_1}} \right] +$$

$$+ \sum_{m=1}^{n-2} w_p^{(m-1)} \left[\sum_{k=0}^{n-m} \binom{n-k}{n-m-k} P_{k-1} \frac{y_{h_1}^{(n-m-k)}}{y_{h_1}} \right]$$

$$= w_p^{(n-1)} \left[\binom{0}{0} P_{0-1} \frac{y_{h_1}^{(0)}}{y_{h_1}} \right] +$$

$$+ w_p^{(n-1-1)} \left[\sum_{k=0}^{1} \binom{n-k}{1-k} P_{k-1} \frac{y_{h_1}^{(1-k)}}{y_{h_1}} \right] +$$

$$+ \sum_{m=1}^{n-2} w_p^{(m-1)} \left[\sum_{k=0}^{n-m} \binom{n-k}{n-m-k} P_{k-1} \frac{y_{h_1}^{(n-m-k)}}{y_{h_1}} \right]$$

$$= w_p^{(n-1)} P_{-1} +$$

$$+ w_p^{(n-1-1)} \left[\binom{n-0}{1-0} P_{0-1} \frac{y_{h_1}^{(1-0)}}{y_{h_1}} + \binom{n-1}{1-1} P_{1-1} \frac{y_{h_1}^{(1-1)}}{y_{h_1}} \right] +$$

$$+ \sum_{m=1}^{n-2} w_p^{(m-1)} \left[\sum_{k=0}^{n-m} \binom{n-k}{n-m-k} P_{k-1} \frac{y_{h_1}^{(n-m-k)}}{y_{h_1}} \right]$$

$$= w_p^{(n-1)} P_{-1} +$$

$$+ w_p^{(n-2)} \left[nP_{-1} \frac{y_{h_1}^{(1)}}{y_{h_1}} + \binom{n-1}{0} P_0 \frac{y_{h_1}^{(0)}}{y_{h_1}} \right] +$$

$$+ \sum_{m=1}^{n-2} w_p^{(m-1)} \left[\sum_{k=0}^{n-m} \binom{n-k}{n-m-k} P_{k-1} \frac{y_{h_1}^{(n-m-k)}}{y_{h_1}} \right]$$

$$= w_p^{(n-1)} + w_p^{(n-2)} \left[n \frac{y_{h_1}'}{y_{h_1}} + P_0 \right] +$$

$$+ \sum_{m=1}^{n-2} w_p^{(m-1)} \left[\sum_{k=0}^{n-m} \binom{n-k}{n-m-k} P_{k-1} \frac{y_{h_1}^{(n-m-k)}}{y_{h_1}} \right]$$

Now, above are formulas for particular solutions of 2nd and 3rd order.
Only the $(n-2)-th$ order term shows up each of the formulas, and

$$y_p = y_{h_1} v_p \Rightarrow w_p = v'_p = \left(\frac{y_p}{y_{h_1}}\right)'$$

As before, $w_\mu = v'_\mu = \left(\frac{y_{h_\mu}}{y_{h_1}}\right)'$, $(y_{h_\mu} \neq y_{h_1})$ is a homogeneous $(n-1)-th$ order solution to:

$$w_\mu^{(n-1)} + w_\mu^{(n-2)}\left[n\frac{y'_{h_1}}{y_{h_1}} + P_0\right] +$$

$$+ \sum_{m=1}^{n-2} w_\mu^{(m-1)}\left[\sum_{k=0}^{n-m} \binom{n-k}{n-m-k} P_{k-1}\frac{y_{h_1}^{(n-m-k)}}{y_{h_1}}\right] = 0;$$

with corresponding inhomogeneous $(n-1)-th$ order:

$$w_p^{(n-1)} + w_p^{(n-2)}\left[n\frac{y'_{h_1}}{y_{h_1}} + P_0\right] +$$

$$+ \sum_{m=1}^{n-2} w_p^{(m-1)}\left[\sum_{k=0}^{n-m} \binom{n-k}{n-m-k} P_{k-1}\frac{y_{h_1}^{(n-m-k)}}{y_{h_1}}\right] = W\frac{1}{y_{h_1}}.$$

where: $y_p = y_{h_1} v_p \Rightarrow w_p = v'_p = \left(\frac{y_p}{y_{h_1}}\right)'.$

[Note that the coefficients of the corresponding w's are the same between the   ]
[two equations, and that the $(n-2)-th$ order terms are both: $n\frac{y'_{h_1}}{y_{h_1}} + P_0$.   ]
[Thus, homogeneous solutions w_μ's correspond to inhomogeneous solutions w_p's]

So,

$$y_p = y_{h_1} \int w_p dx$$

If $n = 4$ the equations in w_μ & w_p are of order: $n-1 = 3$ with solution
given above.
So, for $n = 4$:

$$\left(\frac{y_p}{y_{h_1}}\right)' = v'_p = w_p =$$

$$= w_{h_1} \int\left[\left(\frac{w_{h_2}}{w_{h_1}}\right)'\left\{\int\frac{1}{\left[\left(\frac{w_{h_2}}{w_{h_1}}\right)'\right]^2}\frac{1}{w_{h_1}^3}\frac{1}{y_{h_1}^4} \times\right.\right.$$

$$\times \left(\int\left[\left(W\frac{1}{y_{h_1}}\frac{1}{w_{h_1}}\frac{1}{\left(\frac{w_{h_2}}{w_{h_1}}\right)'}\right)\right]\left[\left(\frac{w_{h_2}}{w_{h_1}}\right)'\right]^2 w_{h_1}^3 y_{h_1}^4 e^{\int P_0 dx} dx\right) e^{-\int P_0 dx} dx\left.\left.\right\}\right]dx,$$

where: w_{h_1} & w_{h_2} are solutions to the homogeneous:

$$w_\mu''' + w_\mu''\left[n\frac{y'_{h_1}}{y_{h_1}} + P_0\right] + w_\mu\left[\sum_{k=0}^{4-1}\binom{4-k}{4-1-k}P_{k-1}\frac{y_{h_1}^{(4-1-k)}}{y_{h_1}}\right] = 0.$$

but as noted above:

$$w_\mu = v'_\mu = \left(\frac{y_{h_\mu}}{y_{h_1}}\right)', (y_{h_\mu} \neq y_{h_1}).$$

are homogeneous solutions of this ODE.

So, substituting:

$$y_p = y_{h_1} \int \left[\left(\frac{y_{h_2}}{y_{h_1}}\right)' \int \left[\left(\frac{\left(\frac{y_{h_3}}{y_{h_1}}\right)'}{\left(\frac{y_{h_2}}{y_{h_1}}\right)'}\right)' \left\{ \int \frac{1}{\left[\left(\frac{\left(\frac{y_{h_3}}{y_{h_1}}\right)'}{\left(\frac{y_{h_2}}{y_{h_1}}\right)'}\right)'\right]^2} \frac{1}{\left(\left(\frac{y_{h_2}}{y_{h_1}}\right)'\right)^3} \frac{1}{y_{h_1}^4} \times \right.\right.$$

$$\times \left(\int \left[\left(W\frac{1}{y_{h_1}}\right)\frac{1}{\left(\frac{y_{h_2}}{y_{h_1}}\right)'} \frac{1}{\left(\frac{\left(\frac{y_{h_3}}{y_{h_1}}\right)'}{\left(\frac{y_{h_2}}{y_{h_1}}\right)'}\right)'}\right] \left[\left(\frac{\left(\frac{y_{h_3}}{y_{h_1}}\right)'}{\left(\frac{y_{h_2}}{y_{h_1}}\right)'}\right)'\right]^2 \left(\left(\frac{y_{h_2}}{y_{h_1}}\right)'\right)^3 y_{h_1}^4 e^{\int P_0 dx} dx \right) \times$$

$$\left. \left. \times e^{-\int P_0 dx} dx \right\} \right] dx \right] dx$$

And, so, finally:

$$y_p = y_{h_1} \int \left[\left(\frac{y_{h_2}}{y_{h_1}}\right)' \int \left[\left(\frac{\left(\frac{y_{h_3}}{y_{h_1}}\right)'}{\left(\frac{y_{h_2}}{y_{h_1}}\right)'}\right)' \left\{ \int \frac{1}{\left[\left(\frac{\left(\frac{y_{h_3}}{y_{h_1}}\right)'}{\left(\frac{y_{h_2}}{y_{h_1}}\right)'}\right)'\right]^2} \frac{1}{\left(\left(\frac{y_{h_2}}{y_{h_1}}\right)'\right)^3} \frac{1}{y_{h_1}^4} \times \right.\right.$$

$$\times \left(\int \left[\left(W\frac{1}{y_{h_1}}\right)\frac{1}{\left(\frac{y_{h_2}}{y_{h_1}}\right)'} \frac{1}{\left(\frac{\left(\frac{y_{h_3}}{y_{h_1}}\right)'}{\left(\frac{y_{h_2}}{y_{h_1}}\right)'}\right)'}\right] \left[\left(\frac{\left(\frac{y_{h_3}}{y_{h_1}}\right)'}{\left(\frac{y_{h_2}}{y_{h_1}}\right)'}\right)'\right]^2 \left(\left(\frac{y_{h_2}}{y_{h_1}}\right)'\right)^3 y_{h_1}^4 e^{\int P_0 dx} dx \right) \times$$

$$\left. \left. \times e^{-\int P_0 dx} dx \right\} \right] dx \right] dx$$

or alternatively:

14

$$y_p = y_{h_1} \int \left[\left(\frac{y_{h_2}}{y_{h_1}}\right)' \int \left[\frac{\left(\frac{y_{h_3}}{y_{h_1}}\right)'}{\left(\frac{y_{h_2}}{y_{h_1}}\right)'} \right)' \right\} \left\{ \int \frac{1}{y_{h_1}^4} \frac{1}{\left(\left(\frac{y_{h_2}}{y_{h_1}}\right)'\right)^3} \frac{1}{\left[\left(\frac{\left(\frac{y_{h_3}}{y_{h_1}}\right)'}{\left(\frac{y_{h_2}}{y_{h_1}}\right)'}\right)'\right]^2} \times \right.$$

$$\times \left(\int \left[\left(W \frac{1}{y_{h_1}}\right) \frac{1}{\left(\frac{y_{h_2}}{y_{h_1}}\right)'} \frac{1}{\left(\frac{\left(\frac{y_{h_3}}{y_{h_1}}\right)'}{\left(\frac{y_{h_2}}{y_{h_1}}\right)'}\right)'} \right] y_{h_1}^4 \left(\left(\frac{y_{h_2}}{y_{h_1}}\right)'\right)^3 \left[\left(\frac{\left(\frac{y_{h_3}}{y_{h_1}}\right)'}{\left(\frac{y_{h_2}}{y_{h_1}}\right)'}\right)'\right]^2 e^{\int P_0 dx} dx \right) \times$$

$$\times \left. e^{-\int P_0 dx} dx \right\} \left] dx \right] dx,$$

is a particular solution for any inhomogeneous linear fourth order ordinary differential equation.

Example 1, consider from [5], Exercise 4.4, 41:

$$y^{(4)} - 2y''' + y'' = e^x + 1$$
$$P_0 = -2 \ , \ W = e^x + 1$$
$$m_1 = 0 \ , \ m_2 = 0 \ , \ m_3 = 1 \ , \ m_4 = 1$$
$$\Rightarrow y_{h_1} = e^{0x} = 1 \ , \ y_{h_2} = xe^{0x} = x \ , \ y_{h_3} = e^x$$
$$\Rightarrow \frac{y_{h_2}}{y_{h_1}} = x \Rightarrow \left(\frac{y_{h_2}}{y_{h_1}}\right)' = 1 \ , \ \frac{y_{h_3}}{y_{h_1}} = e^x \Rightarrow \left(\frac{y_{h_3}}{y_{h_1}}\right)' = e^x \ ,$$
$$\left(\frac{\left(\frac{y_{h_3}}{y_{h_1}}\right)'}{\left(\frac{y_{h_2}}{y_{h_1}}\right)'}\right)' = e^x$$

$$\Rightarrow y_p = (1) \int \left[(1) \int \left[e^x \left\{ \int [e^x]^{-2} [(1)]^{-3} [(1)]^{-4} \times \right. \right. \right.$$
$$\times \left. \left. \left. \left(\int (e^x + 1)\left(\tfrac{1}{1}\right)\left(\tfrac{1}{1}\right)\left(\tfrac{1}{e^x}\right)[e^x]^2[1]^3[1]^4 e^{-2x} dx \right) e^{2x} dx \right\} \right] dx \right] dx$$
$$= \int \left[\int \left[e^x \left\{ \int e^{-2x} \left(\int (e^x + 1)e^x e^{-2x} dx \right) e^{2x} dx \right\} \right] dx \right] dx$$
$$= \int \left[\int \left[e^x \left\{ \int \left(\int (e^x + 1)e^{-x} dx \right) dx \right\} \right] dx \right] dx$$
$$= \int \left[\int \left[e^x \left\{ \int \left(\int (1 + e^{-x}) dx \right) dx \right\} \right] dx \right] dx$$
$$= \int \left[\int \left[e^x \left\{ \int (x - e^{-x}) dx \right\} \right] dx \right] dx$$
$$= \int \left[\int \left[e^x \left(\frac{x^2}{2} + e^{-x} \right) \right] dx \right] dx$$
$$= \int \left[\int \left(\frac{x^2}{2} e^x + 1 \right) dx \right] dx$$

$$= \int \left[\tfrac{1}{2}e^x(x^2 - 2x + 2) + x\right]dx$$

$$= \tfrac{1}{2}e^x(x^2 - 2x + 2) - e^x(x - 1) + e^x + \tfrac{x^2}{2}$$

$$= \tfrac{1}{2}e^x x^2 - xe^x + e^x - xe^x + e^x + e^x + \tfrac{x^2}{2}$$

$$= \tfrac{1}{2}e^x x^2 - 2xe^x + 3e^x + \tfrac{x^2}{2}$$

Since both 0 & 1 are double roots both middle terms are homogeneous solutions, so may be neglected, and the result is the same as noted in [5].

Example 2, consider from [5], Exercise 4.4, 43:

$$16y^{(4)} - y = e^{\frac{1}{2}x} \implies y_p = \tfrac{1}{8}xe^{\frac{1}{2}x}$$

The formula is defined for the coefficient of the highest order term of 1 , so putting the problem in this form:

$$y^{(4)} - \tfrac{1}{16}y = \tfrac{1}{16}e^{\frac{1}{2}x} \implies y_p = \tfrac{1}{8}e^{\frac{1}{2}x}$$

$$P_0 = 0 \ , \quad W = \tfrac{1}{16}e^{\frac{1}{2}x}$$

$$m \in \left\{-\tfrac{1}{2}, \tfrac{1}{2}, \tfrac{i}{2}, -\tfrac{i}{2}\right\}$$

$$y_{h_1} = e^{-\frac{1}{2}x} \ , \quad y_{h_2} = e^{\frac{1}{2}x} \ , \quad y_{h_3} = e^{\frac{1}{2}x}$$

$$\implies \tfrac{y_{h_2}}{y_{h_1}} = e^x \implies \left(\tfrac{y_{h_2}}{y_{h_1}}\right)' = e^x \ ,$$

$$\tfrac{y_{h_3}}{y_{h_1}} = e^{\frac{1}{2}(i+1)x} \implies \left(\tfrac{y_{h_3}}{y_{h_1}}\right)' = \tfrac{1}{2}(i + 1)e^{\frac{1}{2}(i+1)x} \ ,$$

$$\left(\frac{\left(\tfrac{y_{h_3}}{y_{h_1}}\right)'}{\left(\tfrac{y_{h_2}}{y_{h_1}}\right)'}\right)' = \left(\tfrac{1}{2}(i + 1)e^{\frac{1}{2}(i+1)x - x}\right)' = -\tfrac{1}{2}e^{\frac{1}{2}(i-1)x}$$

$$\implies y_p = e^{-\frac{1}{2}x} \int \left[e^x \int \left[\left(-\tfrac{1}{2}e^{\frac{1}{2}(i-1)x}\right)\left\{\int \left[-\tfrac{1}{2}e^{\frac{1}{2}(i-1)x}\right]^{-2}\left[e^x\right]^{-3}\left[e^{-\frac{1}{2}x}\right]^{-4} \times \right. \right. \right.$$

$$\times \left(\int \left(\tfrac{1}{16}e^{\frac{1}{2}x}\right)\left(\tfrac{1}{e^{-\frac{1}{2}x}}\right)\left(\tfrac{1}{e^x}\right)\left(\frac{1}{-\tfrac{1}{2}e^{\frac{1}{2}(i-1)x}}\right)\left[-\tfrac{1}{2}e^{\frac{1}{2}(i-1)x}\right]^2\left[e^x\right]^3\left[e^{-\frac{1}{2}x}\right]^4 e^0 dx\right) \times$$

$$\times e^0 dx\bigg\}\bigg]dx\bigg]dx$$

$$= e^{-\frac{1}{2}x} \int \left[e^x \int \left[\left(e^{\frac{1}{2}(i-1)x}\right) \times \right.\right.$$

$$\times \left\{\int \left[e^{\frac{1}{2}(i-1)x}\right]^{-2}\left[e^x\right]^{-3}\left[e^{-\frac{1}{2}x}\right]^{-4}\left(\tfrac{1}{16}\frac{e^{\frac{1}{2}(i+1)x}}{\tfrac{1}{2}(i + 1)}\right)dx\right\}\bigg]dx\bigg]dx$$

$$= e^{-\frac{1}{2}x} \int \left[e^x \int \left[\left(e^{\frac{1}{2}(i-1)x}\right)\left\{\int \left(\tfrac{1}{16}\frac{e^{\frac{1}{2}(1-i)x}}{\tfrac{1}{2}(1 + i)}\right)dx\right\}\right]dx\right]dx$$

$$= e^{-\frac{1}{2}x} \int \left[e^x \int \left[\left(e^{\frac{1}{2}(i-1)x}\right)\left(\tfrac{1}{16}\frac{e^{\frac{1}{2}(1-i)x}}{\tfrac{1}{4}(1 + i)(1 - i)}\right)\right]dx\right]dx$$

$$= e^{-\frac{1}{2}x} \int \left[e^x \int \left[\tfrac{1}{16}2e^{\frac{1}{2}(i-1)x}e^{\frac{1}{2}(1-i)x}\right]dx\right]dx = e^{-\frac{1}{2}x} \int \left[e^x \int \tfrac{1}{8}dx\right]dx$$

$$= \tfrac{1}{8}e^{-\tfrac{1}{2}x}\int xe^x dx = \tfrac{1}{8}e^{-\tfrac{1}{2}x}e^x(x-1) = \tfrac{1}{8}e^{\tfrac{1}{2}x}(x-1)$$

Since $e^{\tfrac{1}{2}x}$ is a homogeneous solution, this is a particular solution, in agreement with [5].

As was begun in [8], the arbitrary particular solution formula may be obtained starting with a general *n-th* order approach along the same lines as the general reduction of order analysis.

Let: $P_{-1} \equiv 1$,
and:
$$\sum_{m=0}^{n} P_{n-1-m}y^{(m)} = W$$

$$\text{is a } n-th \text{ order ODE.}$$

Let $y = y_0 v$
$$\Rightarrow y^{(m)} = \sum_{k=0}^{m} \binom{m}{k} v^{(m-k)}y_0^{(k)} = \sum_{k=0}^{m} \binom{m}{k} v^{(m)}y_0^{(m-k)}.$$

$\therefore$

$$\sum_{m=0}^{n} P_{n-1-m}y^{(m)} = \sum_{m=0}^{n} P_{n-1-m}\left[\sum_{k=0}^{m} \binom{m}{k} v^{(m-k)}y_0^{(k)}\right]$$
$$= \sum_{m=0}^{n} y_0^{(m)}\left[\sum_{k=m}^{n} \binom{k}{m} P_{m-1-k}v^{(k)}\right]$$
$$= \sum_{m=0}^{n} v^{(m)}\left[\sum_{k=0}^{n-m} \binom{n-k}{n-m-k} P_{k-1}y_0^{(n-m-k)}\right].$$

So:

$$\sum_{m=0}^{n} P_{n-1-m}y^{(m)} = \sum_{k=0}^{n} \binom{n-k}{n-k} P_{k-1}y^{(n-k)}$$
$$= \sum_{m=0}^{n} v^{(m)}\left[\sum_{k=0}^{n-m} \binom{n-k}{n-m-k} P_{k-1}y_0^{(n-m-k)}\right]$$
$$= v^{(0)}\left[\sum_{k=0}^{n-0} \binom{n-k}{n-0-k} P_{k-1}y_0^{(n-0-k)}\right] +$$
$$+ \sum_{m=1}^{n} v^{(m)}\left[\sum_{k=0}^{n-m} \binom{n-k}{n-m-k} P_{k-1}y_0^{(n-m-k)}\right]$$
$$= v^{(0)}\left[\sum_{k=0}^{n} P_{k-1}y_0^{(n-k)}\right] +$$
$$+ \sum_{m=1}^{n} v^{(m)}\left[\sum_{k=0}^{n-m} \binom{n-k}{n-m-k} P_{k-1}y_0^{(n-m-k)}\right]$$

Then, for y_0 such that: $\sum_{k=0}^{n} P_{k-1}y_0^{(n-k)} = 0$:

$$\sum_{m=0}^{n} P_{n-1-m} y^{(m)} = \sum_{m=1}^{n} v^{(m)} \left[\sum_{k=0}^{n-m} \binom{n-k}{n-m-k} P_{k-1} y_0^{(n-m-k)} \right]$$

$v' = w :$

$$\Rightarrow W = \sum_{m=1}^{n} w^{(m-1)} \left[\sum_{k=0}^{n-m} \binom{n-k}{n-m-k} P_{k-1} y_0^{(n-m-k)} \right]$$

As noted earlier, taking the ordered pair $(W, y_0) = \left(0, y_{h_1} = e^{\int s \, dx} \Rightarrow s = \dfrac{y'_{h_1}}{y_{h_1}} \right)$:

$$\Rightarrow 0 = \sum_{m=1}^{n} w_\mu^{(m-1)} \left[\sum_{k=0}^{n-m} \binom{n-k}{n-m-k} P_{k-1} y_{h_1}^{(n-m-k)} \right]$$

$$= \sum_{m=1}^{n} w_\mu^{(m-1)} \left[\sum_{k=0}^{n-m} \binom{n-k}{n-m-k} P_{k-1} \frac{y_{h_1}^{(n-m-k)}}{y_{h_1}} \right]$$

$$= w_\mu^{(n-1)} \left[\sum_{k=0}^{n-n} \binom{n-k}{n-n-k} P_{k-1} \frac{y_{h_1}^{(n-n-k)}}{y_{h_1}} \right] +$$

$$+ w_\mu^{(n-1-1)} \left[\sum_{k=0}^{n-(n-1)} \binom{n-k}{n-(n-1)-k} P_{k-1} \frac{y_{h_1}^{(n-(n-1)-k)}}{y_{h_1}} \right] +$$

$$+ \sum_{m=1}^{n-2} w_\mu^{(m-1)} \left[\sum_{k=0}^{n-m} \binom{n-k}{n-m-k} P_{k-1} \frac{y_{h_1}^{(n-m-k)}}{y_{h_1}} \right]$$

$$= w_\mu^{(n-1)} \left[\binom{0}{0} P_{0-1} \frac{y_{h_1}^{(0)}}{y_{h_1}} \right] +$$

$$+ w_\mu^{(n-1-1)} \left[\sum_{k=0}^{1} \binom{n-k}{1-k} P_{k-1} \frac{y_{h_1}^{(1-k)}}{y_{h_1}} \right] +$$

$$+ \sum_{m=1}^{n-2} w_\mu^{(m-1)} \left[\sum_{k=0}^{n-m} \binom{n-k}{n-m-k} P_{k-1} \frac{y_{h_1}^{(n-m-k)}}{y_{h_1}} \right]$$

$$= w_\mu^{(n-1)} P_{-1} +$$

$$+ w_\mu^{(n-1-1)} \left[\binom{n-0}{1-0} P_{0-1} \frac{y_{h_1}^{(1-0)}}{y_{h_1}} + \binom{n-1}{1-1} P_{1-1} \frac{y_{h_1}^{(1-1)}}{y_{h_1}} \right] +$$

$$+ \sum_{m=1}^{n-2} w_\mu^{(m-1)} \left[\sum_{k=0}^{n-m} \binom{n-k}{n-m-k} P_{k-1} \frac{y_{h_1}^{(n-m-k)}}{y_{h_1}} \right]$$

$$= w_\mu^{(n-1)} P_{-1} +$$

$$+ w_\mu^{(n-2)} \left[n P_{-1} \frac{y_{h_1}^{(1)}}{y_{h_1}} + \binom{n-1}{0} P_0 \frac{y_{h_1}^{(0)}}{y_{h_1}} \right] +$$

$$+ \sum_{m=1}^{n-2} w_\mu^{(m-1)} \left[\sum_{k=0}^{n-m} \binom{n-k}{n-m-k} P_{k-1} \frac{y_{h_1}^{(n-m-k)}}{y_{h_1}} \right]$$

$$= w_\mu^{(n-1)} + w_\mu^{(n-2)} \left[n \frac{y'_{h_1}}{y_{h_1}} + P_0 \right] +$$

$$+ \sum_{m=1}^{n-2} w_\mu^{(m-1)} \left[\sum_{k=0}^{n-m} \binom{n-k}{n-m-k} P_{k-1} \frac{y_{h_1}^{(n-m-k)}}{y_{h_1}} \right]$$

This is a homogeneous $(n-1)-th$ order linear ODE in

$$w_\mu = v'_\mu = \left(\frac{y_{h_\mu}}{y_{h_1}}\right)', (y_{h_\mu} \neq y_{h_1}).$$

And, now, taking the ordered pair $(W, y_0) = (W, y_p)$:

$$\Rightarrow W = \sum_{m=1}^{n} w_p^{(m-1)} \left[\sum_{k=0}^{n-m} \binom{n-k}{n-m-k} P_{k-1} y_{h_1}^{(n-m-k)} \right]$$

$$\Rightarrow W\frac{1}{y_{h_1}} = \sum_{m=1}^{n} w_p^{(m-1)} \left[\sum_{k=0}^{n-m} \binom{n-k}{n-m-k} P_{k-1} \frac{y_{h_1}^{(n-m-k)}}{y_{h_1}} \right]$$

$$= w_p^{(n-1)} \left[\sum_{k=0}^{n-n} \binom{n-k}{n-n-k} P_{k-1} \frac{y_{h_1}^{(n-n-k)}}{y_{h_1}} \right] +$$

$$+ w_p^{(n-1-1)} \left[\sum_{k=0}^{n-(n-1)} \binom{n-k}{n-(n-1)-k} P_{k-1} \frac{y_{h_1}^{(n-(n-1)-k)}}{y_{h_1}} \right] +$$

$$+ \sum_{m=1}^{n-2} w_p^{(m-1)} \left[\sum_{k=0}^{n-m} \binom{n-k}{n-m-k} P_{k-1} \frac{y_{h_1}^{(n-m-k)}}{y_{h_1}} \right]$$

$$= w_p^{(n-1)} \left[\binom{0}{0} P_{0-1} \frac{y_{h_1}^{(0)}}{y_{h_1}} \right] +$$

$$+ w_p^{(n-1-1)} \left[\sum_{k=0}^{1} \binom{n-k}{1-k} P_{k-1} \frac{y_{h_1}^{(1-k)}}{y_{h_1}} \right] +$$

$$+ \sum_{m=1}^{n-2} w_p^{(m-1)} \left[\sum_{k=0}^{n-m} \binom{n-k}{n-m-k} P_{k-1} \frac{y_{h_1}^{(n-m-k)}}{y_{h_1}} \right]$$

$$= w_p^{(n-1)} P_{-1} +$$

$$+ w_p^{(n-1-1)} \left[\binom{n-0}{1-0} P_{0-1} \frac{y_{h_1}^{(1-0)}}{y_{h_1}} + \binom{n-1}{1-1} P_{1-1} \frac{y_{h_1}^{(1-1)}}{y_{h_1}} \right] +$$

$$+ \sum_{m=1}^{n-2} w_p^{(m-1)} \left[\sum_{k=0}^{n-m} \binom{n-k}{n-m-k} P_{k-1} \frac{y_{h_1}^{(n-m-k)}}{y_{h_1}} \right]$$

$$= w_p^{(n-1)} P_{-1} +$$

$$+ w_p^{(n-2)} \left[nP_{-1} \frac{y_{h_1}^{(1)}}{y_{h_1}} + \binom{n-1}{0} P_0 \frac{y_{h_1}^{(0)}}{y_{h_1}} \right] +$$

$$+ \sum_{m=1}^{n-2} w_p^{(m-1)} \left[\sum_{k=0}^{n-m} \binom{n-k}{n-m-k} P_{k-1} \frac{y_{h_1}^{(n-m-k)}}{y_{h_1}} \right]$$

$$= w_p^{(n-1)} + w_p^{(n-2)} \left[n\frac{y'_{h_1}}{y_{h_1}} + P_0 \right] +$$

$$+ \sum_{m=1}^{n-2} w_p^{(m-1)} \left[\sum_{k=0}^{n-m} \binom{n-k}{n-m-k} P_{k-1} \frac{y_{h_1}^{(n-m-k)}}{y_{h_1}} \right]$$

Now, above are formulas for particular
solutions of 2nd and 3rd order.
Only the $(n-2)-th$ order term shows

up each of the formulas, and
$$y_p = y_{h_1} v_p \implies w_p = v'_p = \left(\frac{y_p}{y_{h_1}}\right)'$$

As before, $w_\mu = v'_\mu = \left(\frac{y_{h_\mu}}{y_{h_1}}\right)'$, $(y_{h_\mu} \neq y_{h_1})$
is a homogeneous $(n-1) - th$
order solution to:

$$w_\mu^{(n-1)} + w_\mu^{(n-2)}\left[n\frac{y'_{h_1}}{y_{h_1}} + P_0 \right] +$$

$$+ \sum_{m=1}^{n-2} w_\mu^{(m-1)}\left[\sum_{k=0}^{n-m} \binom{n-k}{n-m-k} P_{k-1}\frac{y_{h_1}^{(n-m-k)}}{y_{h_1}} \right] = 0;$$

with corresponding inhomogeneous
$(n-1) - th$ order:

$$w_p^{(n-1)} + w_p^{(n-2)}\left[n\frac{y'_{h_1}}{y_{h_1}} + P_0 \right] +$$

$$+ \sum_{m=1}^{n-2} w_p^{(m-1)}\left[\sum_{k=0}^{n-m} \binom{n-k}{n-m-k} P_{k-1}\frac{y_{h_1}^{(n-m-k)}}{y_{h_1}} \right] = W\frac{1}{y_{h_1}}.$$

where: $y_p = y_{h_1} v_p \implies w_p = v'_p = \left(\frac{y_p}{y_{h_1}}\right)'.$

[Note that the coefficients of the]
[corresponding w's are the   ]
[same between the   ]
[two equations, and that the   ]
[$(n-2) - th$ order terms are   ]

[both: $n\frac{y'_{h_1}}{y_{h_1}} + P_0$.   ]
[Thus, homogeneous solutions   ]
[w_μ correspond to   ]
[inhomogeneous solutions w_p]

For arbitrary order, we still have:
$$y_p = y_{h_1} \int w_p dx$$
and the equations in w_μ & w_p are
of order: $n-1$ with solution given above.
$$\left(\frac{y_p}{y_{h_1}}\right)' = v'_p = w_p$$
Remembering from above: $w_\mu = v'_\mu = \left(\frac{y_{h_\mu}}{y_{h_1}}\right)'$, $(y_{h_\mu} \neq y_{h_1})$,
let: $w_{h_\mu} = \left(\frac{y_{h_\mu}}{y_{h_1}}\right)'$, $(1 < \mu < n)$
I trust the algorithm/process is clear, and

induction proof is now elementary, but the
final formula may still be unclear.

First,
Define the following stack functional $\Sigma(n,f,g)$ recursively, by:

$$g(f(n)) \equiv \begin{cases} \Sigma(3,f,g) & , (n = 2) \\[2ex] \overbrace{g^{-1}(g^{-1}(\cdots}^{n-3\ of\ these} g^{-1}(\Sigma(n+1))\ \overbrace{\Sigma(n,f,g))\cdots)}^{n-3\ of\ these}\ \Sigma(3,f,g) & , (n > 2) \end{cases}$$

So:

$$\Sigma(3,f,g) = g(f(2))$$
$$g(f(3)) = g^{-1}(\Sigma(4,f,g))\Sigma(3,f,g)$$
$$\Rightarrow \frac{g(f(3))}{g(f(2))} = \frac{g(f(3))}{\Sigma(3,f,g)} = g^{-1}(\Sigma(4,f,g))$$
$$\Rightarrow \Sigma(4,f,g) = g\left(\frac{g(f(3))}{g(f(2))}\right)$$
$$g(f(4)) = g^{-1}(g^{-1}(\Sigma(5,f,g))\Sigma(4,f,g))\Sigma(3,f,g)$$
$$\Rightarrow \frac{g(f(4))}{g(f(2))} = \frac{g(f(3))}{\Sigma(3,f,g)} = g^{-1}(g^{-1}(\Sigma(5,f,g))\Sigma(4,f,g))$$
$$\Rightarrow g\left(\frac{g(f(4))}{g(f(2))}\right) = g^{-1}(\Sigma(5,f,g))\Sigma(4,f,g)$$
$$\Rightarrow \frac{g\left(\frac{g(f(4))}{g(f(2))}\right)}{g\left(\frac{g(f(3))}{g(f(2))}\right)} = \frac{g\left(\frac{g(f(4))}{g(f(2))}\right)}{\Sigma(4,f,g)} = g^{-1}(\Sigma(5,f,g))$$
$$\Rightarrow \Sigma(5,f,g) = g\left(\frac{g\left(\frac{g(f(4))}{g(f(2))}\right)}{g\left(\frac{g(f(3))}{g(f(2))}\right)}\right)$$

and so on.
So, if g is the ordinary derivative, and:
$$f(\mu) = \frac{y h_\mu}{y h_1}\ ,$$
and define: $\Sigma\left(2, \frac{y h_\mu}{y h_1}, \frac{d}{dx}\right) \equiv y h_1$;
then the preceding formulas may be
written:
order 3:

$$y_p = \Sigma\left(2, \frac{y h_\mu}{y h_1}, \frac{d}{dx}\right)\int\Sigma\left(3, \frac{y h_\mu}{y h_1}, \frac{d}{dx}\right) \times$$
$$\times \left\{\int\left[\left[\Sigma\left(2, \frac{y h_\mu}{y h_1}, \frac{d}{dx}\right)\right]^{-3}\Sigma\left(3, \frac{y h_\mu}{y h_1}, \frac{d}{dx}\right)\right]^{-2} \times$$
$$\times \left(\int W\left[\Sigma\left(2, \frac{y h_\mu}{y h_1}, \frac{d}{dx}\right)\Sigma\left(3, \frac{y h_\mu}{y h_1}, \frac{d}{dx}\right)\right]^{-1} \times$$

$$\times \left[\Sigma\left(2, \frac{y_{h_\mu}}{y_{h_1}}, \frac{d}{dx}\right)\right]^3 \left[\Sigma\left(3, \frac{y_{h_\mu}}{y_{h_1}}, \frac{d}{dx}\right)\right]^2 e^{\int P_0 dx} dx\right) e^{-\int P_0 dx} dx \bigg\} dx$$

order 4:

$$y_p = \Sigma\left(2, \frac{y_{h_\mu}}{y_{h_1}}, \frac{d}{dx}\right) \int \left[\Sigma\left(3, \frac{y_{h_\mu}}{y_{h_1}}, \frac{d}{dx}\right) \int \left[\Sigma\left(4, \frac{y_{h_\mu}}{y_{h_1}}, \frac{d}{dx}\right) \times \right.\right.$$

$$\times \bigg\{ \int \left[\Sigma\left(2, \frac{y_{h_\mu}}{y_{h_1}}, \frac{d}{dx}\right)\right]^{-4} \left[\Sigma\left(3, \frac{y_{h_\mu}}{y_{h_1}}, \frac{d}{dx}\right)\right]^{-3} \left[\Sigma\left(4, \frac{y_{h_\mu}}{y_{h_1}}, \frac{d}{dx}\right)\right]^{-2} \times$$

$$\times \left(\int W \left[\Sigma\left(2, \frac{y_{h_\mu}}{y_{h_1}}, \frac{d}{dx}\right) \Sigma\left(3, \frac{y_{h_\mu}}{y_{h_1}}, \frac{d}{dx}\right) \Sigma\left(4, \frac{y_{h_\mu}}{y_{h_1}}, \frac{d}{dx}\right)\right]^{-1} \times$$

$$\times \left[\Sigma\left(2, \frac{y_{h_\mu}}{y_{h_1}}, \frac{d}{dx}\right)\right]^{-4} \left[\Sigma\left(3, \frac{y_{h_\mu}}{y_{h_1}}, \frac{d}{dx}\right)\right]^{-3} \left[\Sigma\left(4, \frac{y_{h_\mu}}{y_{h_1}}, \frac{d}{dx}\right)\right]^{-2} \times$$

$$\times e^{\int P_0 dx} dx\right) e^{-\int P_0 dx} dx \bigg\} \bigg] dx \bigg] dx$$

At last, the final formula is clear.

$$y_p = \Sigma\left(2, \frac{y_{h_\mu}}{y_{h_1}}, \frac{d}{dx}\right) \overbrace{\int \left[\Sigma\left(3, \frac{y_{h_\mu}}{y_{h_1}}, \frac{d}{dx}\right)\cdots\int \left[\Sigma\left(n, \frac{y_{h_\mu}}{y_{h_1}}, \frac{d}{dx}\right)\right.}^{n-2 \text{ of these}} \times$$

$$\times \bigg\{ \int \prod_{k=2}^{n} \left[\Sigma\left(k, \frac{y_{h_\mu}}{y_{h_1}}, \frac{d}{dx}\right)\right]^{k-2-n} \times$$

$$\times \left(\int W \left[\prod_{k=2}^{n} \left[\Sigma\left(k, \frac{y_{h_\mu}}{y_{h_1}}, \frac{d}{dx}\right)\right]\right]^{-1} \prod_{k=2}^{n} \left[\Sigma\left(k, \frac{y_{h_\mu}}{y_{h_1}}, \frac{d}{dx}\right)\right]^{n+2-k} \times$$

$$\times e^{\int P_0 dx} dx\right) e^{-\int P_0 dx} dx \bigg\} \overbrace{\bigg] dx \bigg] dx \bigg] dx}^{n-2 \text{ of these}};$$

is a particular solution for any inhomogeneous linear n–th order ordinary differential equation.

For the record:

$$\Sigma\left(2, \frac{y_{h_\mu}}{y_{h_1}}, \frac{d}{dx}\right) \equiv y_{h_1} \ , \quad \Sigma\left(3, \frac{y_{h_\mu}}{y_{h_1}}, \frac{d}{dx}\right) \equiv \left(\frac{y_{h_2}}{\Sigma\left(2, \frac{y_{h_\mu}}{y_{h_1}}, \frac{d}{dx}\right)}\right)'$$

$$\Sigma\left(4, \frac{y_{h_\mu}}{y_{h_1}}, \frac{d}{dx}\right) \equiv \left(\frac{\left(\frac{y_{h_3}}{\Sigma\left(2, \frac{y_{h_\mu}}{y_{h_1}}, \frac{d}{dx}\right)}\right)'}{\Sigma\left(3, \frac{y_{h_\mu}}{y_{h_1}}, \frac{d}{dx}\right)}\right)'$$

$$\Sigma\left(5, \frac{y_{h_\mu}}{y_{h_1}}, \frac{d}{dx}\right) \equiv \left(\frac{\left(\left(\dfrac{\dfrac{y_{h_4}}{\Sigma\left(2, \frac{y_{h_\mu}}{y_{h_1}}, \frac{d}{dx}\right)}}{\Sigma\left(3, \frac{y_{h_\mu}}{y_{h_1}}, \frac{d}{dx}\right)}\right)'\right)'}{\Sigma\left(4, \frac{y_{h_\mu}}{y_{h_1}}, \frac{d}{dx}\right)}\right)'$$

$$\Sigma\left(6, \frac{y_{h_\mu}}{y_{h_1}}, \frac{d}{dx}\right) \equiv \left(\frac{\left(\frac{\left(\frac{\left(\dfrac{y_{h_5}}{\Sigma\left(2, \frac{y_{h_\mu}}{y_{h_1}}, \frac{d}{dx}\right)}\right)'}{\Sigma\left(3, \frac{y_{h_\mu}}{y_{h_1}}, \frac{d}{dx}\right)}\right)'}{\Sigma\left(4, \frac{y_{h_\mu}}{y_{h_1}}, \frac{d}{dx}\right)}\right)'}{\Sigma\left(5, \frac{y_{h_\mu}}{y_{h_1}}, \frac{d}{dx}\right)}\right)'$$

So, alternatively, a function might be defined:

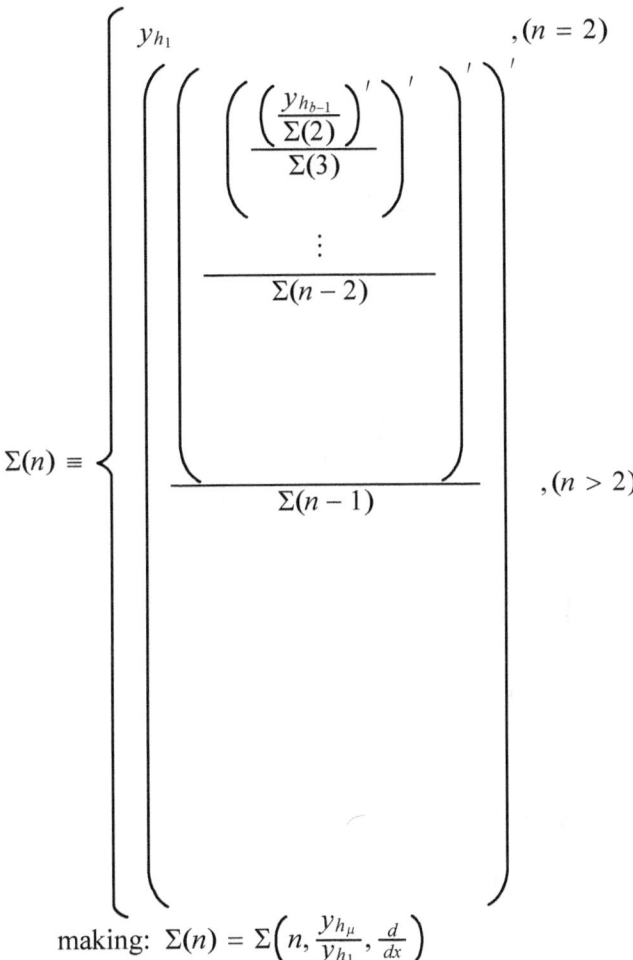

$$\Sigma(n) \equiv \begin{cases} y_{h_1} & ,(n = 2) \\[2em] \cfrac{\left(\left(\left(\cfrac{\left(\cfrac{y_{h_{b-1}}}{\Sigma(2)}\right)'}{\Sigma(3)}\right)' \atop \vdots \right)}{\Sigma(n-2)}\right)'\right)'}{\Sigma(n-1)} & ,(n > 2) \end{cases}$$

making: $\Sigma(n) = \Sigma\left(n, \dfrac{y_{h_\mu}}{y_{h_1}}, \dfrac{d}{dx}\right)$

As you can see; either way, the terms become unwieldy large for placing on a page, which is why they were defined initially as they were. Written this way, eases hand computation for lower orders, and speeds computation by instrumentation at any order.

Fifth and sixth order homogeneous linear constant coefficients ODE may easily be constructed by multiplying monomials together to make auxiliary equations to produce examples with known homogeneous solutions.
Resulting particular solution candidates may then be plugged in for verification.

Example 1 (5th order):
$m \in \{\pm 1, \pm 2, 3\}$
$(m + 1)(m - 1)(m + 2)(m - 2)(m + 3) = 0$

$$(m^2 - 1)(m^2 - 4)(m + 3) = 0$$
$$(m^4 - 5m^2 + 4)(m + 3) = 0$$
$$m^5 + 3m^4 - 5m^3 - 15m^2 + 4m + 12 = 0$$
$$\Rightarrow y^{(5)} + 3y^{(4)} - 5y''' - 15y'' + 4y' + 12y = 0$$
$$\Rightarrow y_h \in \{e^x, e^{-x}, e^{2x}, e^{-2x}, e^{3x}\} \;,\; P_0 = 3$$

Pick a W not the same as one of these: $W = 5e^{7x}$
$$\Rightarrow y^{(5)} + 3y^{(4)} - 5y''' - 15y'' + 4y' + 12y = 5e^{7x}$$
$$y_{h_1} = e^{-x} \;,\; y_{h_2} = e^x \;,\; y_{h_3} = e^{-2x} \;,$$
$$y_{h_4} = e^{2x} \;,\; y_{h_5} = e^{3x}$$

$$\Rightarrow \Sigma\left(2, \frac{y_{h_\mu}}{y_{h_1}}, \frac{d}{dx}\right) = y_{h_1} = e^{-x}$$

$$\Rightarrow \Sigma\left(3, \frac{y_{h_\mu}}{y_{h_1}}, \frac{d}{dx}\right) = \left(\frac{y_{h_2}}{y_{h_1}}\right)' = 2e^{2x}$$

$$\Rightarrow \Sigma\left(4, \frac{y_{h_\mu}}{y_{h_1}}, \frac{d}{dx}\right) = \left(\frac{\left(\frac{y_{h_3}}{y_{h_1}}\right)'}{\Sigma\left(3, \frac{y_{h_\mu}}{y_{h_1}}, \frac{d}{dx}\right)}\right)' = \left(\frac{-e^{-x}}{2e^{2x}}\right)'$$

$$= -\frac{1}{2}\left(e^{-3x}\right)' = \frac{3}{2}e^{-3x}$$

$$\left(\frac{\left(\dfrac{y_{h_4}}{\Sigma\left(2, \frac{y_{h_\mu}}{y_{h_1}}, \frac{d}{dx}\right)}\right)'}{\Sigma\left(3, \frac{y_{h_\mu}}{y_{h_1}}, \frac{d}{dx}\right)}\right)' = \left(\frac{\left(\frac{e^{2x}}{e^{-x}}\right)'}{2e^{2x}}\right)' = \left(\frac{(e^{3x})'}{2e^{2x}}\right)'$$

$$= \left(\frac{3e^{3x}}{2e^{2x}}\right)' = \frac{3}{2}(e^x)' = \frac{3}{2}e^x$$

$$\Rightarrow \Sigma\left(5, \frac{y_{h_\mu}}{y_{h_1}}, \frac{d}{dx}\right) = \left(\frac{\frac{3}{2}e^x}{\frac{3}{2}e^{-3x}}\right)' = (e^{4x})' = 4e^{4x}$$

So:
$$y_p = \Sigma\left(2, \frac{y_{h_\mu}}{y_{h_1}}, \frac{d}{dx}\right) \int \left[\Sigma\left(3, \frac{y_{h_\mu}}{y_{h_1}}, \frac{d}{dx}\right) \int \left[\Sigma\left(4, \frac{y_{h_\mu}}{y_{h_1}}, \frac{d}{dx}\right) \int \left[\Sigma\left(5, \frac{y_{h_\mu}}{y_{h_1}}, \frac{d}{dx}\right) \times \right. \right. \right.$$
$$\times \left\{ \int \left[\Sigma\left(2, \frac{y_{h_\mu}}{y_{h_1}}, \frac{d}{dx}\right)\right]^{2-2-5} \left[\Sigma\left(3, \frac{y_{h_\mu}}{y_{h_1}}, \frac{d}{dx}\right)\right]^{3-2-5} \left[\Sigma\left(4, \frac{y_{h_\mu}}{y_{h_1}}, \frac{d}{dx}\right)\right]^{4-2-5} \left[\Sigma\left(5, \frac{y_{h_\mu}}{y_{h_1}}, \frac{d}{dx}\right) \right.$$
$$\times \left(\int W\left[\Sigma\left(2, \frac{y_{h_\mu}}{y_{h_1}}, \frac{d}{dx}\right)\Sigma\left(3, \frac{y_{h_\mu}}{y_{h_1}}, \frac{d}{dx}\right)\Sigma\left(4, \frac{y_{h_\mu}}{y_{h_1}}, \frac{d}{dx}\right)\Sigma\left(5, \frac{y_{h_\mu}}{y_{h_1}}, \frac{d}{dx}\right)\right]^{-1} \times \right.$$
$$\times \left[\Sigma\left(2, \frac{y_{h_\mu}}{y_{h_1}}, \frac{d}{dx}\right)\right]^{5+2-2} \left[\Sigma\left(3, \frac{y_{h_\mu}}{y_{h_1}}, \frac{d}{dx}\right)\right]^{5+2-3} \left[\Sigma\left(4, \frac{y_{h_\mu}}{y_{h_1}}, \frac{d}{dx}\right)\right]^{5+2-4} \left[\Sigma\left(5, \frac{y_{h_\mu}}{y_{h_1}}, \frac{d}{dx}\right)\right.$$
$$\times e^{\int P_0 dx} dx \Big) e^{-\int P_0 dx} dx \Big] dx \Big] dx \Big] dx;$$

$$= e^{-x} \int \left[2e^{2x} \int \left[\frac{3}{2}e^{-3x} \int \left[4e^{4x} \times \right. \right. \right.$$
$$\times \left\{ \int [e^{-x}]^{2-2-5} [2e^{2x}]^{3-2-5} \left[\frac{3}{2}e^{-3x}\right]^{4-2-5} [4e^{4x}]^{5-2-5} \times \right.$$
$$\times \left(\int (5e^{7x})\left[(e^{-x})(2e^{2x})\left(\frac{3}{2}e^{-3x}\right)(4e^{4x})\right]^{-1} \times \right.$$
$$\times [e^{-x}]^{5+2-2} [2e^{2x}]^{5+2-3} \left[\frac{3}{2}e^{-3x}\right]^{5+2-4} [4e^{4x}]^{5+2-5} e^{3x} dx\Big) e^{-3x} dx \Big\} \Big] dx \Big] dx \Big] dx;$$

26

$$= 12e^{-x} \int \left[e^{2x} \int \left[e^{-3x} \int \left[e^{4x} \times \right. \right. \right.$$

$$\times \left\{ \int e^{5x} 2^{-4} e^{-8x} \left(\tfrac{3}{2} \right)^{-3} e^{9x} 4^{-2} e^{-8x} \left(\int (5e^{7x}) [12e^{2x}]^{-1} \times \right. \right.$$

$$\left. \times e^{-5x} 2^4 e^{8x} \left(\tfrac{3}{2} \right)^3 e^{-9x} 4^2 e^{8x} e^{3x} dx \right) e^{-3x} dx \left. \right\} \left. \right] dx \left. \right] dx \left. \right] dx;$$

$$= e^{-x} \int \left[e^{2x} \int \left[e^{-3x} \int \left[e^{4x} \left\{ \int e^{-2x} \left(\int (5e^{7x}) [e^{2x}]^{-1} \times \right. \right. \right. \right. \right.$$

$$\left. \times e^{2x} e^{3x} dx \right) e^{-3x} dx \left. \right\} \left. \right] dx \left. \right] dx \left. \right] dx;$$

$$= e^{-x} \int \left[e^{2x} \int \left[e^{-3x} \int \left[e^{4x} \left\{ \int e^{-2x} \left(\int 5e^{10x} dx \right) e^{-3x} dx \right\} \right] dx \right] dx \right] dx;$$

$$= e^{-x} \int \left[e^{2x} \int \left[e^{-3x} \int \left[e^{4x} \left\{ \int e^{-5x} \left(\tfrac{5e^{10x}}{10} \right) dx \right\} \right] dx \right] dx \right] dx;$$

$$= e^{-x} \int \left[e^{2x} \int \left[e^{-3x} \int \left[e^{4x} \left\{ \tfrac{1}{2} \int e^{5x} dx \right\} \right] dx \right] dx \right] dx;$$

$$= e^{-x} \int \left[e^{2x} \int \left[e^{-3x} \int \left[e^{4x} \left(\tfrac{1}{2} \tfrac{e^{5x}}{5} \right) \right] dx \right] dx \right] dx;$$

$$= e^{-x} \int \left[e^{2x} \int \left[e^{-3x} \left(\tfrac{1}{10} \int e^{9x} dx \right) \right] dx \right] dx;$$

$$= e^{-x} \int \left[e^{2x} \int \left[e^{-3x} \left(\tfrac{1}{10} \tfrac{e^{9x}}{9} \right) \right] dx \right] dx;$$

$$= e^{-x} \int \left[e^{2x} \left(\tfrac{1}{90} \int e^{6x} dx \right) \right] dx;$$

$$= e^{-x} \int \left[e^{2x} \left(\tfrac{1}{90} \tfrac{e^{6x}}{6} \right) \right] dx;$$

$$= \tfrac{1}{540} e^{-x} \int e^{8x} dx;$$

$$= \tfrac{1}{540} e^{-x} \tfrac{e^{8x}}{8} = \tfrac{1}{4320} e^{7x}$$

And, verifying:

$$y_p' = \tfrac{7}{4320} e^{7x} , \quad y_p'' = \tfrac{49}{4320} e^{7x} , \quad y_p''' = \tfrac{343}{4320} e^{7x} ,$$

$$y_p^{(4)} = \tfrac{2401}{4320} e^{7x} , \quad y_p^{(5)} = \tfrac{16807}{4320} e^{7x}$$

$$\Rightarrow y_p^{(5)} + 3y_p^{(4)} - 5y_p''' - 15y_p'' + 4y_p' + 12y_p =$$

$$= e^{7x} \left[\tfrac{16807}{4320} + 3 \left(\tfrac{2401}{4320} \right) - 5 \left(\tfrac{343}{4320} \right) - 15 \left(\tfrac{49}{4320} \right) + 4 \left(\tfrac{7}{4320} \right) + 12 \left(\tfrac{1}{4320} \right) \right]$$

$$= \tfrac{1}{4320} e^{7x} [16807 + 3(2401) - 5(343) - 15(49) + 4(7) + 12(1)]$$

$$= \tfrac{1}{4320} e^{7x} [16807 + 7203 - 1715 - 735 + 28 + 12]$$

$$= \tfrac{1}{4320} e^{7x} [16807 + 7203 - 1715 - 735 + 28 + 12] = \tfrac{21600}{4320} e^{7x}$$

$$= 5e^{7x}$$

Wow.

Example 2 (6th order):

$$m \in \{ \pm 1, \pm 2, \pm 3 \}$$

$$(m^4 - 5m^2 + 4)(m^2 - 9) = 0$$

$$m^6 - 14m^4 + 49m^2 - 36 = 0$$

$$\Rightarrow y^{(6)} - 14y^{(4)} + 49y'' - 36y = 0$$

$$\Rightarrow y_h \in \{ e^x, e^{-x}, e^{2x}, e^{-2x}, e^{3x}, e^{-3x} \} , \quad P_0 = 0$$

Pick a W the same as one of these: $W = 7e^{3x}$

$$\Rightarrow y^{(6)} - 14y^{(4)} + 49y'' - 36y = 7e^{3x}$$

$$y_{h_1} = e^{-x} , \quad y_{h_2} = e^x , \quad y_{h_3} = e^{-2x} ,$$

$$y_{h_4} = e^{2x} \ , \ y_{h_5} = e^{3x} \ , \ y_{h_6} = e^{-3x}$$

The $\Sigma\left(m, \dfrac{y_{h_\mu}}{y_{h_1}}, \dfrac{d}{dx}\right)$ are all the same as example 1, except for the last one.

$$\left(\dfrac{\left(\dfrac{\left(\dfrac{\Sigma\left(2, \dfrac{y_{h_\mu}}{y_{h_1}}, \dfrac{d}{dx}\right)}{y_{h_5}}\right)'}{\Sigma\left(3, \dfrac{y_{h_\mu}}{y_{h_1}}, \dfrac{d}{dx}\right)}\right)'}{\Sigma\left(4, \dfrac{y_{h_\mu}}{y_{h_1}}, \dfrac{d}{dx}\right)}\right)' = \left(\dfrac{\left(\dfrac{\left(\dfrac{e^{3x}}{e^{-x}}\right)'}{2e^{2x}}\right)'}{\tfrac{3}{2}e^{-3x}}\right)'$$

$$= \left(\dfrac{\left(\dfrac{(e^{4x})'}{2e^{2x}}\right)'}{\tfrac{3}{2}e^{-3x}}\right)' = \left(\dfrac{\tfrac{1}{2}\left(\dfrac{4e^{4x}}{e^{2x}}\right)'}{\tfrac{3}{2}e^{-3x}}\right)'$$

$$= \left(\dfrac{2(e^{2x})'}{3e^{-3x}}\right)' = \left(\dfrac{2(2e^{2x})}{3e^{-3x}}\right)' = \tfrac{4}{3}(e^{5x})' = \tfrac{20}{3}e^{5x}$$

$$\Rightarrow \Sigma\left(6, \dfrac{y_{h_\mu}}{y_{h_1}}, \dfrac{d}{dx}\right) = \left(\dfrac{\tfrac{20}{3}e^{5x}}{4e^{4x}}\right)' = \tfrac{5}{3}e^{x}$$

So:

$$y_p = \Sigma\left(2, \dfrac{y_{h_\mu}}{y_{h_1}}, \dfrac{d}{dx}\right) \int \left[\Sigma\left(3, \dfrac{y_{h_\mu}}{y_{h_1}}, \dfrac{d}{dx}\right) \int \left[\Sigma\left(4, \dfrac{y_{h_\mu}}{y_{h_1}}, \dfrac{d}{dx}\right) \times\right.\right.$$

$$\times \int \left[\Sigma\left(5, \dfrac{y_{h_\mu}}{y_{h_1}}, \dfrac{d}{dx}\right) \int \left[\Sigma\left(6, \dfrac{y_{h_\mu}}{y_{h_1}}, \dfrac{d}{dx}\right) \times\right.\right.$$

$$\times \left\{ \int \left[\Sigma\left(2, \dfrac{y_{h_\mu}}{y_{h_1}}, \dfrac{d}{dx}\right)\right]^{2-2-6} \left[\Sigma\left(3, \dfrac{y_{h_\mu}}{y_{h_1}}, \dfrac{d}{dx}\right)\right]^{3-2-6} \left[\Sigma\left(4, \dfrac{y_{h_\mu}}{y_{h_1}}, \dfrac{d}{dx}\right)\right]^{4-2-6} \times \right.$$

$$\times \left[\Sigma\left(5, \dfrac{y_{h_\mu}}{y_{h_1}}, \dfrac{d}{dx}\right)\right]^{5-2-6} \left[\Sigma\left(6, \dfrac{y_{h_\mu}}{y_{h_1}}, \dfrac{d}{dx}\right)\right]^{6-2-6} \times$$

$$\times \left(\int W \left[\Sigma\left(2, \dfrac{y_{h_\mu}}{y_{h_1}}, \dfrac{d}{dx}\right)\Sigma\left(3, \dfrac{y_{h_\mu}}{y_{h_1}}, \dfrac{d}{dx}\right)\Sigma\left(4, \dfrac{y_{h_\mu}}{y_{h_1}}, \dfrac{d}{dx}\right)\Sigma\left(5, \dfrac{y_{h_\mu}}{y_{h_1}}, \dfrac{d}{dx}\right)\Sigma\left(6, \dfrac{y_{h_\mu}}{y_{h_1}}, \dfrac{d}{dx}\right)\right]^{-1}$$

$$\times \left[\Sigma\left(2, \dfrac{y_{h_\mu}}{y_{h_1}}, \dfrac{d}{dx}\right)\right]^{6+2-2} \left[\Sigma\left(3, \dfrac{y_{h_\mu}}{y_{h_1}}, \dfrac{d}{dx}\right)\right]^{6+2-3} \left[\Sigma\left(4, \dfrac{y_{h_\mu}}{y_{h_1}}, \dfrac{d}{dx}\right)\right]^{6+2-4} \times$$

$$\times \left[\Sigma\left(5, \dfrac{y_{h_\mu}}{y_{h_1}}, \dfrac{d}{dx}\right)\right]^{6+2-5} \left[\Sigma\left(6, \dfrac{y_{h_\mu}}{y_{h_1}}, \dfrac{d}{dx}\right)\right]^{6+2-6} \times$$

$$\times e^{\int P_0 dx} dx \left)e^{-\int P_0 dx} dx\right\} \, dx \right] dx \right] dx \right] dx;$$

$$= e^{-x} \int \left[2e^{2x} \int \left[\tfrac{3}{2}e^{-3x} \int \left[4e^{4x} \int \left[\tfrac{5}{3}e^{x} \times\right.\right.\right.\right.$$

28

$$\times\left\{\int[e^{-x}]^{2-2-6}[2e^{2x}]^{3-2-6}\left[\tfrac{3}{2}e^{-3x}\right]^{4-2-6}[4e^{4x}]^{5-2-6}\left[\tfrac{5}{3}e^{x}\right]^{6-2-6}\times\right.$$

$$\times\left(\int(7e^{3x})\left[(e^{-x})(2e^{2x})(\tfrac{3}{2}e^{-3x})(4e^{4x})(\tfrac{5}{3}e^{x})\right]^{-1}\times\right.$$

$$\times[e^{-x}]^{6+2-2}[2e^{2x}]^{6+2-3}\left[\tfrac{3}{2}e^{-3x}\right]^{6+2-4}[4e^{4x}]^{6+2-5}\left[\tfrac{5}{3}e^{x}\right]^{6+2-6}e^{0}dx\Big)e^{0}dx\Big\}\Big]dx\Big]dx\Big]dx$$

$$=e^{-x}\int\left[2e^{2x}\int\left[\tfrac{3}{2}e^{-3x}\int\left[4e^{4x}\int\left[\tfrac{5}{3}e^{x}\times\right.\right.\right.\right.$$

$$\times\left\{\int[e^{-x}]^{-6}[2e^{2x}]^{-5}\left[\tfrac{3}{2}e^{-3x}\right]^{-4}[4e^{4x}]^{-3}\left[\tfrac{5}{3}e^{x}\right]^{-2}\times\right.$$

$$\times\left(\int(7e^{3x})\left[(e^{-x})(2e^{2x})(\tfrac{3}{2}e^{-3x})(4e^{4x})(\tfrac{5}{3}e^{x})\right]^{-1}\times\right.$$

$$\times[e^{-x}]^{6}[2e^{2x}]^{5}\left[\tfrac{3}{2}e^{-3x}\right]^{4}[4e^{4x}]^{3}\left[\tfrac{5}{3}e^{x}\right]^{2}e^{0}dx\Big)e^{0}dx\Big\}\Big]dx\Big]dx\Big]dx\Big]dx;$$

$$=e^{-x}\int\left[e^{2x}\int\left[e^{-3x}\int\left[e^{4x}\int\left[e^{x}\left\{\int e^{6x}e^{-10x}e^{12x}e^{-12x}e^{-2x}\times\right.\right.\right.\right.\right.$$

$$\times\left(\int(7e^{3x})[e^{-x}e^{2x}e^{-3x}e^{4x}e^{x}]^{-1}e^{-6x}e^{10x}e^{-12x}e^{12x}e^{2x}dx\Big)dx\right\}\Big]dx\Big]dx\Big]dx\Big]dx;$$

$$=e^{-x}\int\left[e^{2x}\int\left[e^{-3x}\int\left[e^{4x}\int\left[e^{x}\left\{\int e^{-6x}\times\right.\right.\right.\right.\right.$$

$$\times\left(\int(7e^{3x})e^{-3x}e^{6x}dx\Big)dx\right\}\Big]dx\Big]dx\Big]dx\Big]dx;$$

$$=e^{-x}\int\left[e^{2x}\int\left[e^{-3x}\int\left[e^{4x}\int\left[e^{x}\left\{\int e^{-6x}\left(\int 7e^{6x}dx\right)dx\right\}\right]dx\right]dx\right]dx\right]dx;$$

$$=e^{-x}\int\left[e^{2x}\int\left[e^{-3x}\int\left[e^{4x}\int\left[e^{x}\left\{\int e^{-6x}\left(7\tfrac{e^{6x}}{6}\right)dx\right\}\right]dx\right]dx\right]dx\right]dx;$$

$$=e^{-x}\int\left[e^{2x}\int\left[e^{-3x}\int\left[e^{4x}\int\left[e^{x}\left\{\tfrac{7}{6}\int dx\right\}\right]dx\right]dx\right]dx\right]dx;$$

$$=e^{-x}\int\left[e^{2x}\int\left[e^{-3x}\int\left[e^{4x}\left[\tfrac{7}{6}\int xe^{x}dx\right]\right]dx\right]dx\right]dx;$$

$$=e^{-x}\int\left[e^{2x}\int\left[e^{-3x}\left[\tfrac{7}{6}\int\left[e^{4x}e^{x}[x-1]\right]dx\right]\right]dx\right]dx;$$

$$=e^{-x}\int\left[e^{2x}\int\left[e^{-3x}\left[\tfrac{7}{6}\int e^{5x}[x-1]dx\right]\right]dx\right]dx;$$

$$=e^{-x}\int\left[e^{2x}\int\left[e^{-3x}\left[\tfrac{7}{6}e^{5x}\left(\tfrac{1}{25}[5x-1]-1\right)\right]\right]dx\right]dx;$$

$$=\tfrac{7}{6}e^{-x}\int\left[e^{2x}\int e^{2x}\left(\tfrac{1}{5}x-\tfrac{1}{25}-1\right)dx\right]dx;$$

$$=\tfrac{7}{6}e^{-x}\int\left[e^{2x}\left(\tfrac{1}{5}\int xe^{2x}dx+\left[-\tfrac{1}{25}-1\right]\int e^{2x}dx\right)\right]dx;$$

$$=\tfrac{7}{6}e^{-x}\int\left[e^{2x}\left(\tfrac{1}{5}\tfrac{e^{2x}}{4}[2x-1]+\left[-\tfrac{1}{25}-1\right]\tfrac{e^{2x}}{2}\right)\right]dx;$$

$$=\tfrac{7}{6}e^{-x}\int e^{4x}\left(\tfrac{1}{5}\tfrac{1}{4}[2x-1]+\tfrac{1}{2}\left[-\tfrac{1}{25}-1\right]\right)dx;$$

$$=\tfrac{7}{6}e^{-x}\int e^{4x}\left(\tfrac{1}{5}\tfrac{1}{4}[2x-1]dx+\tfrac{1}{2}\left[-\tfrac{1}{25}-1\right]\int e^{4x}dx\right);$$

$$=\tfrac{7}{6}e^{-x}\left(\tfrac{1}{5}\tfrac{1}{4}\left[2\int xe^{4x}dx-\int e^{4x}dx\right]+\tfrac{1}{2}\left[-\tfrac{1}{25}-1\right]\tfrac{e^{4x}}{4}\right);$$

$$=\tfrac{7}{6}e^{-x}\left(\tfrac{1}{5}\tfrac{1}{4}\left[2\int xe^{4x}dx-\tfrac{e^{4x}}{4}\right]+\tfrac{1}{2}\left[-\tfrac{1}{25}-1\right]\tfrac{e^{4x}}{4}\right);$$

$$=\tfrac{7}{6}\left\{\tfrac{1}{5}\tfrac{1}{4}\left[2e^{-x}\tfrac{e^{4x}}{16}[4x-1]-\tfrac{1}{4}e^{3x}\right]+\tfrac{1}{2}\left[-\tfrac{1}{25}-1\right]\tfrac{1}{4}e^{3x}\right\};$$

$$=\tfrac{7}{6}\left\{\tfrac{1}{5}\tfrac{1}{4}\left[2e^{3x}\left[\tfrac{1}{4}x-\tfrac{1}{16}\right]-\tfrac{1}{4}e^{3x}\right]+\tfrac{1}{2}\left[-\tfrac{1}{25}-1\right]\tfrac{1}{4}e^{3x}\right\};$$

Since e^{3x} is a homogeneous solution drop all the only e^{3x} terms.

$$=\tfrac{7}{6}\tfrac{1}{5}\tfrac{1}{4}2\tfrac{1}{4}xe^{3x}=\tfrac{7}{6}\tfrac{1}{5}\tfrac{1}{8}xe^{3x}=\tfrac{7}{240}xe^{3x};$$

And, verifying:
$$y_p' = \tfrac{7}{240}e^{3x}(1+3x) \ , \ y_p'' = \tfrac{7}{240}e^{3x}(6+9x) \ ,$$
$$y_p''' = \tfrac{7}{240}e^{3x}(27+27x) \ , \ y_p^{(4)} = \tfrac{7}{240}e^{3x}(108+81x) \ ,$$
$$y_p^{(5)} = \tfrac{7}{240}e^{3x}(405+243x) \ , \ y_p^{(6)} = \tfrac{7}{240}e^{3x}(1458+729x)$$
$$\Rightarrow y_p^{(6)} - 14y_p^{(4)} + 49y_p'' - 36y_p =$$
$$= \tfrac{7}{240}e^{3x}\Big[x[729 - 14(81) + 49(9) - 36(1)] +$$
$$+ [1458 - 14(108) + 49(6) - 36(0)]\Big]$$
$$= \tfrac{7}{240}e^{3x}\Big[x[729 - 1134 + 441 - 36] +$$
$$+ [1458 - 1512 + 294 - 36(0)]\Big]$$
$$= \tfrac{7}{240}e^{3x}(x[0] + [240]) = 7e^{3x}$$

OMG.

Orders from 2nd to 4th are already completely proven. I have worked out 5th order longhand, also, to understand the general form. Those and the above two examples, are already beyond proof beyond a reasonable doubt.

At this point, the proof by induction is elementary.
As already shown, the formula is true for: $N = 2, 3, 4$.
Assuming it is true for $n = N$:
For $n = N + 1$:
$$y_p = y_{h_1}v_p \Rightarrow w_p = v_p' = \left(\frac{y_p}{y_{h_1}}\right)'$$
And, as above: $w_\mu = v_\mu' = \left(\frac{y_{h_\mu}}{y_{h_1}}\right)'$, $(y_{h_\mu} \neq y_{h_1})$,
is a homogeneous $(n-1) - th$ order solution to:
$$w_\mu^{(n-1)} + w_\mu^{(n-2)}\left[n\frac{y_{h_1}'}{y_{h_1}} + P_0\right] +$$
$$+ \sum_{m=1}^{n-2} w_\mu^{(m-1)}\left[\sum_{k=0}^{n-m}\binom{n-k}{n-m-k}P_{k-1}\frac{y_{h_1}^{(n-m-k)}}{y_{h_1}}\right] = 0;$$
with corresponding inhomogeneous $(n-1) - th$ order:
$$w_p^{(n-1)} + w_p^{(n-2)}\left[n\frac{y_{h_1}'}{y_{h_1}} + P_0\right] +$$

$$+ \sum_{m=1}^{n-2} w_p^{(m-1)} \left[\sum_{k=0}^{n-m} \binom{n-k}{n-m-k} P_{k-1} \frac{y_{h_1}^{(n-m-k)}}{y_{h_1}} \right] = W \frac{1}{y_{h_1}}.$$

let: $w_{h_\mu} = \left(\dfrac{y_{h_\mu}}{y_{h_1}} \right)'$, $(1 < \mu < n)$

For arbitrary order, we still have:

$$y_p = y_{h_1} \int w_p \, dx$$

And:

$$w_p = \Sigma\left(2, \frac{w_{h_\mu}}{w_{h_1}}, \frac{d}{dx}\right) \overbrace{\int \left[\Sigma\left(3, \frac{w_{h_\mu}}{w_{h_1}}, \frac{d}{dx}\right) \cdots \int \left[\Sigma\left(N, \frac{w_{h_\mu}}{w_{h_1}}, \frac{d}{dx}\right)\right.}^{N-2 \text{ of these}} \times$$

$$\times \left\{ \int \prod_{k=2}^{N} \left[\Sigma\left(k, \frac{w_{h_\mu}}{w_{h_1}}, \frac{d}{dx}\right)\right]^{k-2-N} \times \right.$$

$$\times \left(\int \left(W\frac{1}{y_{h_1}}\right) \left[\prod_{k=2}^{N} \left[\Sigma\left(k, \frac{w_{h_\mu}}{w_{h_1}}, \frac{d}{dx}\right)\right]\right]^{-1} \prod_{k=2}^{N}\left[\Sigma\left(k, \frac{w_{h_\mu}}{w_{h_1}}, \frac{d}{dx}\right)\right]^{N+2-k} \times \right.$$

$$\times \left. e^{\int P_0 dx} dx \right) e^{-\int P_0 dx} dx \right\} \overbrace{\left. \right] dx \left. \right] dx \right] dx}^{N-2 \text{ of these}};$$

$$= \Sigma\left(3, \frac{y_{h_\mu}}{y_{h_1}}, \frac{d}{dx}\right) \overbrace{\int \left[\Sigma\left(4, \frac{y_{h_\mu}}{y_{h_1}}, \frac{d}{dx}\right) \cdots \int \left[\Sigma\left(N+1, \frac{y_{h_\mu}}{y_{h_1}}, \frac{d}{dx}\right)\right.}^{N-2 \text{ of these}} \times$$

$$\times \left\{ \int \prod_{k=3}^{N+1} \left[\Sigma\left(k, \frac{y_{h_\mu}}{y_{h_1}}, \frac{d}{dx}\right)\right]^{k-2-(N+1)} \times \right.$$

$$\times \left(\int \left(W\frac{1}{y_{h_1}}\right) \left[\prod_{k=3}^{N+1} \left[\Sigma\left(k, \frac{y_{h_\mu}}{y_{h_1}}, \frac{d}{dx}\right)\right]\right]^{-1} \prod_{k=3}^{N+1}\left[\Sigma\left(k, \frac{y_{h_\mu}}{y_{h_1}}, \frac{d}{dx}\right)\right]^{(N+1)+2-k} \times \right.$$

$$\times \left. e^{\int \left(P_0 + (N+1)\frac{y'_{h_\mu}}{y_{h_1}}\right) dx} dx \right) e^{-\int \left(P_0 + (N+1)\frac{y'_{h_\mu}}{y_{h_1}}\right) dx} dx \right\} \overbrace{\left. \right] dx \left. \right] dx \right] dx}^{N-2 \text{ of these}};$$

$$= \Sigma\left(3, \frac{y_{h_\mu}}{y_{h_1}}, \frac{d}{dx}\right) \overbrace{\int \left[\Sigma\left(4, \frac{y_{h_\mu}}{y_{h_1}}, \frac{d}{dx}\right) \cdots \int \left[\Sigma\left(N+1, \frac{y_{h_\mu}}{y_{h_1}}, \frac{d}{dx}\right)\right.}^{N-2 \text{ of these}} \times$$

$$\times \left\{ \int \prod_{k=2}^{N+1} \left[\Sigma\left(k, \frac{y_{h_\mu}}{y_{h_1}}, \frac{d}{dx}\right)\right]^{k-2-(N+1)} \times \right.$$

$$\times \left(\int W \left[\prod_{k=2}^{N+1} \left[\Sigma\left(k, \frac{y_{h_\mu}}{y_{h_1}}, \frac{d}{dx}\right)\right]\right]^{-1} \prod_{k=2}^{N+1}\left[\Sigma\left(k, \frac{y_{h_\mu}}{y_{h_1}}, \frac{d}{dx}\right)\right]^{(N+1)+2-k} \times \right.$$

$$\times \left. e^{\int P_0 dx} dx \right) e^{-\int P_0 dx} dx \right\} \overbrace{\left. \right] dx \left. \right] dx \right] dx}^{N-2 \text{ of these}}.$$

And, so, for $n = N+1$:

$$\Rightarrow y_p = y_{h_1} \int \left[\Sigma\left(3, \frac{y_{h_\mu}}{y_{h_1}}, \frac{d}{dx}\right) \overbrace{\int \left[\Sigma\left(4, \frac{y_{h_\mu}}{y_{h_1}}, \frac{d}{dx}\right) \cdots \int \left[\Sigma\left(N+1, \frac{y_{h_\mu}}{y_{h_1}}, \frac{d}{dx}\right)\right.}^{N-2 \text{ of these}} \times$$

$$\times \left\{ \int \prod_{k=2}^{N+1} \left[\Sigma\left(k, \frac{y_{h_\mu}}{y_{h_1}}, \frac{d}{dx}\right) \right]^{k-2-(N+1)} \times \right.$$

$$\times \left(\int W \left[\prod_{k=2}^{N+1} \left[\Sigma\left(k, \frac{y_{h_\mu}}{y_{h_1}}, \frac{d}{dx}\right) \right] \right]^{-1} \prod_{k=2}^{N+1} \left[\Sigma\left(k, \frac{y_{h_\mu}}{y_{h_1}}, \frac{d}{dx}\right) \right]^{(N+1)+2-k} \times \right.$$

$$\underbrace{}_{N-2 \text{ of these}}$$

$$\times \left. e^{\int P_0 dx} dx \right) e^{-\int P_0 dx} dx \Big\} \overbrace{}^{} \,]dx \,]dx \,]dx \,]dx;$$

$$\underbrace{}_{N-2 \text{ of these}}$$

$$= \Sigma\left(2, \frac{y_{h_\mu}}{y_{h_1}}, \frac{d}{dx}\right) \int \left[\Sigma\left(3, \frac{y_{h_\mu}}{y_{h_1}}, \frac{d}{dx}\right) \int \left[\Sigma\left(4, \frac{y_{h_\mu}}{y_{h_1}}, \frac{d}{dx}\right) \cdots \int \left[\Sigma\left(N+1, \frac{y_{h_\mu}}{y_{h_1}}, \frac{d}{dx}\right) \times \right. \right. \right.$$

$$\times \left\{ \int \prod_{k=2}^{N+1} \left[\Sigma\left(k, \frac{y_{h_\mu}}{y_{h_1}}, \frac{d}{dx}\right) \right]^{k-2-(N+1)} \times \right.$$

$$\times \left(\int W \left[\prod_{k=2}^{N+1} \left[\Sigma\left(k, \frac{y_{h_\mu}}{y_{h_1}}, \frac{d}{dx}\right) \right] \right]^{-1} \prod_{k=2}^{N+1} \left[\Sigma\left(k, \frac{y_{h_\mu}}{y_{h_1}}, \frac{d}{dx}\right) \right]^{(N+1)+2-k} \times \right.$$

$$\underbrace{}_{(N+1)-2 \text{ of these}}$$

$$\times \left. e^{\int P_0 dx} dx \right) e^{-\int P_0 dx} dx \Big\} \,]dx \,]dx \,]dx \,]dx;$$

$$\underbrace{}_{(N+1)-2 \text{ of these}}$$

$$= \Sigma\left(2, \frac{y_{h_\mu}}{y_{h_1}}, \frac{d}{dx}\right) \int \left[\Sigma\left(3, \frac{y_{h_\mu}}{y_{h_1}}, \frac{d}{dx}\right) \int \left[\Sigma\left(4, \frac{y_{h_\mu}}{y_{h_1}}, \frac{d}{dx}\right) \cdots \int \left[\Sigma\left(N+1, \frac{y_{h_\mu}}{y_{h_1}}, \frac{d}{dx}\right) \times \right. \right. \right.$$

$$\times \left\{ \int \prod_{k=2}^{N+1} \left[\Sigma\left(k, \frac{y_{h_\mu}}{y_{h_1}}, \frac{d}{dx}\right) \right]^{k-2-(N+1)} \times \right.$$

$$\times \left(\int W \left[\prod_{k=2}^{N+1} \left[\Sigma\left(k, \frac{y_{h_\mu}}{y_{h_1}}, \frac{d}{dx}\right) \right] \right]^{-1} \prod_{k=2}^{N+1} \left[\Sigma\left(k, \frac{y_{h_\mu}}{y_{h_1}}, \frac{d}{dx}\right) \right]^{(N+1)+2-k} \times \right.$$

$$\underbrace{}_{(N+1)-2 \text{ of these}}$$

$$\times \left. e^{\int P_0 dx} dx \right) e^{-\int P_0 dx} dx \Big\} \,]dx \,]dx \,]dx \,]dx.$$

$\square$

A Small Table of Particular Solutions For Inhomogeneous Linear Ordinary Differential Equations of Second Order

Another technique from that of [6] uses the reduction of order technique to establish the second order particular solution formula as follows.

Let y_P be a solution to the inhomogeneous ODE:
$$y_p'' + Py_p' + Qy_p = W.$$

Then consider:
$$y_p = v e^{\int s\,dx},$$
where, as above, s satisfies: $Q = -s' - s^2 - sP.$

Differentiating:
$$y_p' = (v' + vs) e^{\int s\,dx},$$
$$y_p'' = (v'' + v's + vs') e^{\int s\,dx} + (v' + vs)s e^{\int s\,dx}.$$

So,
$$e^{\int s\,dx}[v'' + 2v's + vs' + vs^2 + P(v' + vs) + Qv] = W.$$
$$\Rightarrow v'' + 2v's + vs' + vs^2 + P(v' + vs) + Qv = W e^{-\int s\,dx}.$$
$$\Rightarrow v'' + (2s + P)v' + (s' + s^2 + Ps + Q)v = W e^{-\int s\,dx}.$$
$$\Rightarrow v'' + (2s + P)v' = W e^{-\int s\,dx}.$$
$$\Rightarrow e^{-\int (2s+P)dx}\left(v' e^{\int (2s+P)dx}\right)' = W e^{-\int s\,dx}.$$
$$\Rightarrow v = \int e^{-\int (2s+P)dx}\left(\int W e^{-\int s\,dx} e^{\int (2s+P)dx} dx\right)dx.$$
$$\Rightarrow v = \int e^{-\int (2s+P)dx}\left(\int W e^{\int (s+P)dx} dx\right)dx.$$
$$\Rightarrow y_p = e^{\int s\,dx}\int e^{-\int (2s+P)dx}\left(\int W e^{\int (s+P)dx} dx\right)dx.$$

But, since s satisfies: $Q = -s' - s^2 - sP$, then:
$$y_h = e^{\int s\,dx} \Rightarrow y_h'' + P y_h' + Q y_h = 0.$$
Substituting:
$$y_p = y_h \int \frac{1}{y_h^2}\left(\int W y_h e^{\int P\,dx} dx\right) e^{-\int P\,dx} dx,$$
as above.

Still another method of obtaining the second order particular solution formula is presented in [6] as Corollary II.2 .

Tables of particular solutions may be constructed from this formula, just as tables of derivatives and integrals are.

As an example, consider the liner inhomogeneous ODEs with constant coefficients:
$$y'' + Ay' + By = C e^{ax} + D e^{bx}.$$
$$y_h = e^{mx}, \text{ where } m \text{ is a root of the auxiliary equation.}$$
So,
$$y_p = e^{mx}\int e^{-2mx}\left(\int (C e^{ax} + D e^{bx})e^{(m+A)x} dx\right)e^{-Ax} dx.$$
$$\Rightarrow y_p = e^{mx}\int\left(\int C e^{(a+m+A)x} + D e^{(b+m+A)x} dx\right)e^{-(2m+A)x} dx.$$
$$\Rightarrow y_p = e^{mx}\int\left(\frac{C}{a+m+A} e^{(a+m+A)x} + \frac{D}{b+m+A} e^{(b+m+A)x}\right)e^{-(2m+A)x} dx, (a + m + A \neq 0)$$

$$\Rightarrow y_p = e^{mx} \int \left(\frac{C}{a+m+A} e^{(a+m+A-2m-A)x} + \frac{D}{b+m+A} e^{(b+m+A-2m-A)x} \right) dx.$$

$$\Rightarrow y_p = e^{mx} \int \left(\frac{C}{a+m+A} e^{(a-m)x} + \frac{D}{b+m+A} e^{(b-m)x} \right) dx.$$

$$\Rightarrow y_p = e^{mx} \left(\frac{C}{(a+m+A)(a-m)} e^{(a-m)x} + \frac{D}{(b+m+A)(b-m)} e^{(b-m)x} \right),$$

$$(a+m+A \neq 0, a-m \neq 0)$$

$$\Rightarrow y_p = Re^{ax} + Se^{bx},$$

where :
$$R = \begin{cases} \dfrac{C}{(a+m+A)(a-m)} & ,(a+m+A \neq 0, a-m \neq 0) \\[2mm] \dfrac{C}{(a+m+A)} x & ,(a+m+A \neq 0, a-m = 0) \\[2mm] \dfrac{C}{(a-m)}(x+1) & ,(a+m+A = 0, a-m \neq 0) \\[2mm] \dfrac{C}{2} x^2 & ,(a+m+A = 0, a-m = 0) \end{cases}$$

$$S = \begin{cases} \dfrac{D}{(b+m+A)(a-m)} & ,(b+m+A \neq 0, b-m \neq 0) \\[2mm] \dfrac{D}{(b+m+A)} x & ,(b+m+A \neq 0, b-m = 0) \\[2mm] \dfrac{D}{(b-m)}(x+1) & ,(b+m+A = 0, b-m \neq 0) \\[2mm] \dfrac{D}{2} x^2 & ,(b+m+A = 0, b-m = 0) \end{cases}$$

(*note* :
$$a+m+A = 0 \Rightarrow m-a = -A-a-a = -A-2a = -A+2m+2A = 2m+A)$$

I believe it is clear, without going into detail, that:
$$y'' + Ay' + By = \sum_{j=1}^{N} C_j e^{a_j x} \Rightarrow y_p = \sum_{j=1}^{N} S_j e^{a_j x},$$

where : $S_j = \begin{cases} \dfrac{C_j}{(a_j+m+A)(a_j-m)} & ,(a_j+m+A \neq 0, a_j-m \neq 0) \\[2mm] \dfrac{C_j}{(a_j+m+A)} x & ,(a_j+m+A \neq 0, a_j-m = 0) \\[2mm] \dfrac{C_j}{(a_j-m)}(x+1) & ,(a_j+m+A = 0, a_j-m \neq 0) \\[2mm] \dfrac{C_j}{2} x^2 & ,(a_j+m+A = 0, a_j-m = 0) \end{cases}$

Also, since sines and cosines may be written as sums/differences of complex exponentials, this method is particularly useful separating them into their real and imaginary parts. Use the above formula, and recombine them back into the original sines and cosines as desired.

As another example, suppose:

$$y'' + Ay' + By = \sum_{j=1}^{N} C_j e^{a_j x} \sin(\phi_j x) + \sum_{j=1}^{M} D_j e^{b_j x} \cos(\psi_j x) \ , \ \left(\phi_j \, , \, \psi_j \in \mathbb{R} \right)$$

Using the method suggested, above:

$$\Rightarrow y'' + Ay' + By = \sum_{j=1}^{N} C_j e^{a_j x} \left(\tfrac{1}{2i} [e^{i\phi_j x} - e^{-i\phi_j x}] \right) +$$

$$+ \sum_{j=1}^{M} D_j e^{b_j x} \left(\tfrac{1}{2} [e^{i\psi_j x} + e^{-i\psi_j x}] \right)$$

$$= \tfrac{1}{2i} \sum_{j=1}^{N} C_j e^{(a_j + i\phi_j)x} - \tfrac{1}{2i} \sum_{j=1}^{N} C_j e^{(a_j - i\phi_j)x} +$$

$$+ \tfrac{1}{2} \sum_{j=1}^{M} D_j e^{(b_j + i\psi_j)x} + \tfrac{1}{2} \sum_{j=1}^{M} D_j e^{(b_j - i\psi_j)x}$$

$$\Rightarrow y_p = \tfrac{1}{2i} \sum_{j=1}^{N} S_j(C_j, a_j + i\phi_j) e^{(a_j + i\phi_j)x} - \tfrac{1}{2i} \sum_{j=1}^{N} S_j(C_j, a_j - i\phi_j) e^{(a_j - i\phi_j)x}$$

$$+ \tfrac{1}{2} \sum_{j=1}^{M} S_j(D_j, b_j + i\psi_j) e^{(b_j + i\psi_j)x} + \tfrac{1}{2} \sum_{j=1}^{M} S_j(D_j, b_j - i\psi_j) e^{(b_j - i\psi_j)x}$$

$$= \tfrac{1}{2i} \sum_{j=1}^{N} e^{a_j x} \left\{ S_j(C_j, a_j + i\phi_j) e^{i\phi_j x} - S_j(C_j, a_j - i\phi_j) e^{-i\phi_j x} \right\} +$$

$$+ \tfrac{1}{2} \sum_{j=1}^{M} e^{b_j x} \left\{ S_j(D_j, b_j + i\psi_j) e^{i\psi_j x} + S_j(D_j, b_j - i\psi_j) e^{-i\psi_j x} \right\}$$

$$= \tfrac{1}{2i} \sum_{j=1}^{N} e^{a_j x} \left\{ S_j(C_j, a_j + i\phi_j) \left(\tfrac{1}{2} [e^{i\phi_j x} + e^{-i\phi_j x}] + i \tfrac{1}{2i} [e^{i\phi_j x} - e^{-i\phi_j x}] \right) \right\} +$$

$$- \tfrac{1}{2i} \sum_{j=1}^{M} e^{a_j x} \left\{ S_j(C_j, a_j - i\phi_j) \left(\tfrac{1}{2} [e^{i\phi_j x} + e^{-i\phi_j x}] - i \tfrac{1}{2i} [e^{i\phi_j x} - e^{-i\phi_j x}] \right) \right\} +$$

$$+ \tfrac{1}{2} \sum_{j=1}^{M} e^{b_j x} \left\{ S_j(D_j, b_j + i\psi_j) \left(\tfrac{1}{2} [e^{i\psi_j x} + e^{-i\psi_j x}] + i \tfrac{1}{2i} [e^{i\psi_j x} - e^{-i\psi_j x}] \right) \right\} +$$

$$+ \tfrac{1}{2} \sum_{j=1}^{M} e^{b_j x} \left\{ S_j(D_j, b_j - i\psi_j) \left(\tfrac{1}{2} [e^{i\psi_j x} + e^{-i\psi_j x}] - i \tfrac{1}{2i} [e^{i\psi_j x} - e^{-i\psi_j x}] \right) \right\}$$

$$= \tfrac{1}{2i} \sum_{j=1}^{N} e^{a_j x} \left\{ S_j(C_j, a_j + i\phi_j) [\cos(\phi_j x) + i \sin(\phi_j x)] \right\} +$$

$$- \tfrac{1}{2i} \sum_{j=1}^{M} e^{a_j x} \left\{ S_j(C_j, a_j - i\phi_j) [\cos(\phi_j x) - i \sin(\phi_j x)] \right\} +$$

$$+ \tfrac{1}{2} \sum_{j=1}^{M} e^{b_j x} \left\{ S_j(D_j, b_j + i\psi_j) [\cos(\psi_j x) + i \sin(\psi_j x)] \right\} +$$

$$+ \tfrac{1}{2} \sum_{j=1}^{M} e^{b_j x} \left\{ S_j(D_j, b_j - i\psi_j) [\cos(\psi_j x) - i \sin(\psi_j x)] \right\}$$

$$= \tfrac{1}{2i} \sum_{j=1}^{N} e^{a_j x} \left\{ [S_j(C_j, a_j + i\phi_j) - S_j(C_j, a_j - i\phi_j)] \cos(\phi_j x) \right\} +$$

$$+ \frac{1}{2i} \sum_{j=1}^{M} e^{a_j x} \left\{ [S_j(C_j, a_j + i\phi_j) + S_j(C_j, a_j - i\phi_j)]i\sin(\phi_j x) \right\} +$$

$$+ \frac{1}{2} \sum_{j=1}^{M} e^{b_j x} \left\{ [S_j(D_j, b_j + i\psi_j) + S_j(D_j, b_j - i\psi_j)]\cos(\psi_j x) \right\} +$$

$$+ \frac{1}{2} \sum_{j=1}^{M} e^{b_j x} \left\{ [S_j(D_j, b_j + i\psi_j) - S_j(D_j, b_j - i\psi_j)]i\sin(\psi_j x) \right\}$$

$$= \frac{1}{2i} \sum_{j=1}^{N} e^{a_j x} \left\{ [S_j(C_j, a_j + i\phi_j) - S_j(C_j, a_j - i\phi_j)]\cos(\phi_j x) \right\} +$$

$$+ \frac{1}{2} \sum_{j=1}^{M} e^{a_j x} \left\{ [S_j(C_j, a_j + i\phi_j) + S_j(C_j, a_j - i\phi_j)]\sin(\phi_j x) \right\} +$$

$$+ \frac{1}{2} \sum_{j=1}^{M} e^{b_j x} \left\{ [S_j(D_j, b_j + i\psi_j) + S_j(D_j, b_j - i\psi_j)]\cos(\psi_j x) \right\} +$$

$$+ \frac{i}{2} \sum_{j=1}^{M} e^{b_j x} \left\{ [S_j(D_j, b_j + i\psi_j) - S_j(D_j, b_j - i\psi_j)]\sin(\psi_j x) \right\}$$

$$where : S_j(P,k) = \begin{cases} \dfrac{P}{(k^* + m + A)(k^* - m)}, & (k^* + m + A \neq 0, k^* - m \neq 0) \\[2ex] \dfrac{P}{(k^* + m + A)}x & ,(k^* + m + A \neq 0, k^* - m = 0) \\[2ex] \dfrac{P}{(k^* - m)}(x + 1) & ,(k^* + m + A = 0, k^* - m \neq 0) \\[2ex] \dfrac{P}{2}x^2 & ,(k^* + m + A = 0, k^* - m = 0) \\[2ex] & ,\left(k^* = k + n\pi , \ \exists n \in \mathbb{I}\right) \\[1ex] \text{(in practice, the number of } n \text{ to test is small)} \end{cases}$$

Since integrals of polynomials times exponentials may be evaluated via integration by parts (or from integral tables obtained that way) this formula is also an effective way of evaluating particular solutions for these types (and may also be tabularized as above). In fact, the list of types that may be so tabularized may grow exceeding large.

So, as a further example, suppose:
$$y'' + Ay' + By = \sum_{j=1}^{N} C_j x^{b_j} e^{a_j x}$$

Since:
$$\int x^n e^{mx} dx = e^{mx} \sum_{k=0}^{n} \binom{n}{k} \frac{k!}{m^{k+1}} x^{n-k}$$

$$\Rightarrow y_p = e^{mx} \int e^{-2mx} \left(\int \left(\sum_{j=1}^{N} C_j x^{b_j} e^{a_j x} \right) e^{(m+A)x} dx \right) e^{-Ax} dx.$$

$$\Rightarrow y_p = \sum_{j=1}^{N} e^{mx} \int e^{-2mx} \left(\int C_j x^{b_j} e^{(a_j+m+A)x} dx \right) e^{-Ax} dx.$$

$$\Rightarrow y_p = \sum_{j=1}^{N} e^{mx} \int e^{-2mx} \left[C_j e^{(a_j+m+A)x} \sum_{k=0}^{b_j} \binom{b_j}{k} \frac{k!}{(a_j+m+A)^{k+1}} x^{b_j-k} \right] e^{-Ax} dx.$$

$$\Rightarrow y_p = \sum_{j=1}^{N} \sum_{k=0}^{b_j} e^{mx} \int \left[C_j \binom{b_j}{k} \frac{k!}{(a_j+m+A)^{k+1}} x^{b_j-k} e^{(a_j+m+A-A-2m)x} \right] dx.$$

$$\Rightarrow y_p = \sum_{j=1}^{N} \sum_{k=0}^{b_j} C_j \binom{b_j}{k} \frac{k!}{(a_j+m+A)^{k+1}} e^{mx} \int \left[x^{b_j-k} e^{(a_j-m)x} \right] dx.$$

$$\Rightarrow y_p = \sum_{j=1}^{N} \sum_{k=0}^{b_j} C_j \binom{b_j}{k} \frac{k!}{(a_j+m+A)^{k+1}} e^{mx} e^{(a_j-m)x} \sum_{h=0}^{b_j-k} \binom{b_j-k}{h} \frac{h!}{(a_j-m)^{h+1}} x^{b_j-k-h}$$

$$\Rightarrow y_p = \sum_{j=1}^{N} e^{a_j x} \left[C_j \sum_{k=0}^{b_j} \binom{b_j}{k} \frac{k!}{(a_j+m+A)^{k+1}} \sum_{h=0}^{b_j-k} \binom{b_j-k}{h} \frac{h!}{(a_j-m)^{h+1}} x^{b_j-k-h} \right].$$

$$where : b_j \in \mathbb{N}, \forall j$$

I believe it is clear, without going into detail, that:

$$y'' + Ay' + By = \sum_{j=1}^{N} C_j x^{b_j} e^{a_j x} \Rightarrow y_p = \sum_{j=1}^{N} S_j e^{a_j x}, \quad where :$$

$$S_j = \begin{cases} \left[C_j \sum_{k=0}^{b_j} \binom{b_j}{k} \frac{k!}{(a_j+m+A)^{k+1}} \sum_{h=0}^{b_j-k} \binom{b_j-k}{h} \frac{h!}{(a_j-m)^{h+1}} \right], \\ \qquad\qquad\qquad\qquad\qquad\qquad\qquad\qquad (a_j+m+A \neq 0, a_j - m \neq 0) \\[4pt] \left[C_j \sum_{k=0}^{b_j} \binom{b_j}{k} \frac{k!}{(a_j+m+A)^{k+1}(b_j+1-k)} \right] x^{b_j+1-k}, (a_j+m+A \neq 0, a_j - m = 0) \\[4pt] \left[C_j \sum_{k=0}^{b_j+1} \binom{b_j+1}{k} \frac{k!}{(a_j-m)^{k+1}} \right] x^{b_j+1-k} \qquad , (a_j+m+A = 0, a_j - m \neq 0) \\[4pt] \frac{C_j}{(b_j+1)(b_j+2)} x^{b_j+2} \qquad\qquad\qquad , (a_j+m+A = 0, a_j - m = 0) \end{cases}$$

The references, [2], [3], and [5] have been used to verify these results; and reference [6] further contains a pair of verifying worked out example solutions as well as a short proof.

Again, powers of sines and cosines may be written as sums and differences of multiple angle sines and cosines, i.e. sums and differences of complex exponentials, so this formula may be applied immediately, as above.

Clearly, also, this formula is a convenient starting point to prove that the standard forms of a particular solution are as generally presented in introductory ODE texts. As demonstrated above, it is no more computationally extensive than the methods of undetermined coefficients or variation of parameters; and certainly less so than the Wronskian and convolution methods.

Computationally, this method must be much more compact in memory and execute faster; considering the particular solutions may be tabularized as above, and each of the possibilities jumped to and computed immediately - as opposed to the other methods being crunched out the old "mindless" ways.

And, as an added bonus, using the above examples and techniques, this method makes creating text exercizes with solutions virtually effortless.

References

[1]   Kamke, E.; *Differentialgleichungen Lösungsmethoden Und Lösungen,*
      3rd Ed., Chelsea Publishing Company, New York, N. Y.; 1959.

[2]   Nagle, R.K. , & Saff, E.B.; *Fundamentals of Differential Equations and*
      Boundary Value Problems; Addison Wesley Publishing Company, Inc.;
      Reading, MA; 1994.

[3]   Nagle, R.K. , & Saff, E.B., & Snider, A.D.; *Fundamentals of Differential*
      Equations, 5th Ed.; Addison Wesley Longman, Inc.; Reading, MA; 2000.

[4]   Polyanin, Andrei D. & Zaitsev, Valentin F.; *Handbook of Exact Solutions*
      for Ordinary Differential Equations, 2nd. Ed.; Chapman & Hall/CRC;
      New York, NY; 2003.

[5]   Zill, Dennis G.; *A First Course in Differential Equations with Applications,*
      4th Ed.; PWS-KENT Publishing Company; Boston, MA; 1989.

[6]   SciVee: DOI: 10.4016/28294.01 , http://www.scivee.tv/node/28294 ;
http://www.dnatube.com/video/6899/A-Particular-Solutions-Inhomogeneous-2nd-Order-ODE

[7]   Cassano, Claude M.;
http://www.dnatube.com/video/6967/A-Particular-Solutions-Inhomogeneous-3rd-Order-ODE

[8]   Cassano, Claude M.;
http://www.dnatube.com/video/6968/A-Particular-Solutions-Inhomogeneous-4th-Order-ODE

www.ingramcontent.com/pod-product-compliance
Lightning Source LLC
Chambersburg PA
CBHW071551170526
45166CB00004B/1627